Sexual Crime

Sexual Crime is an edited book series devised by a team at SOCAMRU led by Professor Belinda Winder in the Psychology Division at Nottingham Trent University.

It offers original contributions to specific avenues of research within the field of sexual crime with each volume drawing together a review of the literature from across disciplines, including cutting edge research and practice, original material from services and offenders themselves as well as future directions for research and practice.

These volumes will be highly relevant to clinical and forensic psychologists and therapists, offender managers and supervisors, social workers and those working in the field of rehabilitation. They will be a great source of insight for academics, researchers and students in these disciplines as well as criminologists and policy makers.

Nicholas Blagden · Belinda Winder ·
Kerensa Hocken · Rebecca Lievesley ·
Craig A. Harper · Helen Swaby ·
Phil Banyard
Editors

Sexual Crime and Community Reintegration

palgrave
macmillan

Editors
Nicholas Blagden
University of Derby
Derby, UK

Kerensa Hocken
HMP Whatton
Nottingham, UK

Craig A. Harper
Department of Psychology
Nottingham Trent University
Nottingham, UK

Phil Banyard
Department of Psychology
Nottingham Trent University
Nottingham, UK

Belinda Winder
Department of Psychology
Nottingham Trent University
Nottingham, UK

Rebecca Lievesley
Department of Psychology
Nottingham Trent University
Nottingham, UK

Helen Swaby
Department of Psychology
Nottingham Trent University
Nottingham, UK

ISSN 2628-8168 ISSN 2628-8176 (electronic)
Sexual Crime
ISBN 978-3-032-06033-4 ISBN 978-3-032-06034-1 (eBook)
https://doi.org/10.1007/978-3-032-06034-1

Cover credit: com6402/Imazins

This Palgrave Macmillan imprint is published by the registered company Springer Nature Switzerland AG
The registered company address is: Gewerbestrasse 11, 6330 Cham, Switzerland

Preface

This Series: Sexual Crime

This book series will offer original contributions to current books available on this fast growing area of high public interest. Each volume will comprehensively engage with current literature, and make efforts to access unpublished literature and data by key authors in the field. The series will also, by the end of each volume, suggest potential new directions for researchers and practitioners.

These volumes are relevant not just to psychologists, criminologists, social workers, final year undergraduate, postgraduate, doctoral students of all these areas, but to practitioners and the general public with an interest in learning more about the topic. The aim is to create books that are readable, yet firmly anchored in a sound evidence base from both researchers and practitioners. The volumes will include therefore a robust synthesis of the literature, consideration of the theories relevant to each topic, a focus on projects that are relevant to the topic, with a summary of the research and evaluation of these, chapters focusing on the service user voice and a final summary chapter, highlighting future possibilities and directions (as suggested by others in the field or by the authors themselves).

This Volume: Sexual Crime and Community Reintegration

The overwhelming majority (>98%) of individuals in prison for sexual offences are men and most will be released into the community. Proven reoffending, defined as an offence that leads to a conviction or other formal sanction within a 12-month follow-up period after release from custody, for sexual offences has a lower recidivism rate than all other offence types and this includes individuals whose index offences were against adults. Given that there has been an increase in the number of people in prison for a sexual offence conviction, it is notable that holistic and well-rounded support for people when they are released from prison has remained limited. Nearly 25 years ago Maruna contended that "the question of how to improve the process of ex-offender re-entry or reintegration will be among the most important issues facing the country in the next few decades". Yet here we are nearly a quarter of a century later and there are very few support services for those with sexual convictions as they transition into the community, despite the empirical evidence that such support can keep individuals and communities safe. Reintegration and rehabilitation of this group remains challenging due to numerous social, structural and political challenges; employment, finding a place within society and moving forward are all challenging for those with sexual convictions.

In response to this our book hopes to capture the experience of transitioning from prison to the community, the tensions between risk management, public protection and enabling flourishing in those that have been released from prison. In this book Chapter 1 explores a series of processes, both psychological and social, which present persistent barriers to reintegration efforts. It highlights how mortification and mythology are entwined social processes, which shape both the experiences of individuals with sexual offence convictions and the societal responses existing around them in engrained ways. This then creates the context for Chapter 2 which explores how specialist police teams responsible for the Management of Sexual/Violent Offenders (MOSOVOs) practice within a moral tension between public protection and supporting individuals with sexual convictions in the context of desistance. It suggests that successful desistance relies on collaboration in which MOSOVOs are willing to help those with sexual convictions by providing safe, non-stigmatising, and stable interactions. Chapter 3 explores the other side of these experiences by focusing on men with sexual convictions attempting to reintegrate into the community. This chapter draws on interviews with nine men released from prison and focuses on how these men negotiated the "sex offender" label. It argues that restrictions such as being on the 'sex offender' register and carrying the label of 'sex offender' can negatively impact the desistance process and increase stigma for this population. It further emphasises that in many criminal justice professions the "sex offender" label becomes the relational mechanism by which others relate to them. Chapter 4 highlights the realities of resettlement including the stresses around disclosure (and anticipatory disclosure). The chapter highlights that processes such as disclosing a criminal record is one which can cause significant distress due to its perpetual nature, and the lack of social provision to counter encountered stigma compounds the experience of shame. This chapter concludes that there must be stable provision for people who have harmed, and been harmed, to be supported. Chapter 5 builds on the previous chapters by focusing on the transition from prison to the community. It poses the critical question do we want them to succeed or fail? While the chapters in this book upto Chapter 5 have been critical discussions of challenges and responses to them, Chapter 6 explores the lived experiences of those released into the community in their own words and is authored by two service users of the Safer Living Foundation. It highlights, in their words, the role of the Safer Living Centre in the process of reintegration, with the men arguing that such supportive environments play a meaningful role in helping individuals rebuild their lives post-release. Chapter 7 explores themes raised by service-users in Chapter 6 and explores a central aspect in any reintegration into the community for men with sexual convictions i.e. the experiences of living with and navigating licence conditions. This chapter argues that licence conditions can act as a barrier to reintegration when they are not done in consultation with the individual and applied without criticality. The chapter argues that certain conditions e.g. unable to "associate with a known offender", may work counter to the expected outcome. The chapter concludes that licence conditions need more thought and should be done collaboratively. The final chapter, Chapter 8, explores trauma-informed care in a high intensity treatment programme for individuals with histories of adverse childhood events (ACEs). It focuses on trauma in those with sexual convictions and It argues that addressing trauma-related traits can support rehabilitation. Using a case study of an adult male with multiple ACEs, it suggests that trauma-informed care helps manage risk, develop metacognitive skills, improve self-regulation, and support reintegration. Taken together these chapters highlight, for the first time in one volume, in explicit detail the experience of transition from prison to the community, the barriers and challenges to reintegration, but also how we can move forward, how we can help men live meaningful pro-social lives and in doing so fulfil the goal that everyone working in this field wants—to reduce sexual (re) victimisation.

Nottingham, UK Nicholas Blagden
Nottingham, UK Belinda Winder
Nottingham, UK Kerensa Hocken
Nottingham, UK Rebecca Lievesley
Nottingham, UK Craig A. Harper
Lincoln, UK Helen Swaby
Nottingham, UK Phil Banyard

Acknowledgements

This book is dedicated to the Safer Living Foundation, a charity that for 11 years worked to help men transition from prison to the community, offering social, emotional and practical support to help men reintegrate and lead offence-free lives. In the 11 years of being in operation the charity worked with over 300 individuals through a range of supportive projects. Beyond accolades, including winning Charity of the Year in 2019, the legacy of the SLF will be in the crimes that have been prevented through the people the SLF supported and the changes the men made while a part of the SLF. The book is also dedicated to all the men that went through the SLF's services, the men who committed to change and made positive differences in their lives.

We also acknowledge the proof reading assistance of Eve Penford and Jade Mason during the refinement of the chapters.

Contents

List of Tables

1

Mortification and Mythology: Challenges in the Reintegration of People with Sexual Offence Convictions

Christian Perrin and William McGowan

Introduction

This chapter is fundamentally about the challenge of reintegrating people with sexual offence convictions (PwSOC) especially, though not exclusively, following periods of incarceration. Specifically, it explores a series of processes, both psychological and social, which present persistent barriers to reintegration efforts. In doing so, the chapter is divided into four substantive subsections. The first provides some important context concerning the societal tendency, justified or not, to attribute something of a 'special status' to sexual offence convictions. The two sections following this are intended to act as heuristic devices by cataloguing

C. Perrin (✉)
University of Birmingham, Birmingham, UK
e-mail: c.s.perrin@bham.ac.uk

W. McGowan
School of Law and Justice Studies, Liverpool John Moores University, Liverpool, UK

N. Blagden et al. (eds.), *Sexual Crime and Community Reintegration*, Sexual Crime,
https://doi.org/10.1007/978-3-032-06034-1_1

predominantly psychological processes of 'mortification and degradation', on the one hand, and predominantly social and cultural processes of 'mythologising', on the other.

As with most heuristic devices, these mortifications and mythologies are mutually constitutive and overlapping processes, rather than discrete and exclusive. On this basis, readers can decide for themselves as to the empirical utility of categorising complex, messy, and incendiary phenomena (the treatment of PwSOC by both the criminal justice system and the public at large) in this way. Amidst the political noise surrounding debates about 'sex offenders', this chapter argues that there is analytic and moral value in spotlighting certain psychological impacts on individuals with sexual offence convictions—especially from the perspective of community reintegration. Simultaneously, it emphasises the need to distinguish these impacts from the social and cultural tropes that fuel public fears about this category of offending behaviour. The repercussions of these twinned processes—mortification and mythology—on reintegration and prevention efforts are then discussed before some concluding thoughts on this most contentious of topics.

Sexual Offence: An Ecological Niche in the Collective Consciousness

No other crime ignites public fear and condemnation as intensely as sexual offence, particularly that committed against children. Labels such as 'sex offender', 'child molester', and 'paedophile' occupy a distinct ecological niche within society's collective consciousness, commonly provoking levels of disgust difficult to imagine within other contextual frames. This 'sex offender exceptionalism' (Yung, 2011) is not limited to the general public but can pervade professional/practitioner circles, including those working in forensic and clinical settings (Blagden et al., 2013). The stigmatising force of such labels (and the mythological images they invoke) is so powerful that it typically overrides any higher-order critical thinking or nuanced reasoning, leading to entrenched misconceptions about PwSOC.

Accordingly, despite contrasting with research evidence (Fortney et al., 2007) "*the dominant narrative construction, in Western societies, concerning [especially] child sex offenders identifies such individuals as purely male, inherently evil, inhuman, beyond redemption or cure, lower class, and unknown to the victim (who is constructed as female)*" (Gavin, 2005, p.395). As a result, PwSOC are frequently viewed as 'beyond the pale' (Harkins, 2020)—irredeemable predators existing outside the boundaries of moral and social redemption. This narrative framework constitutes much of the landscape, and 'starting point', for those hoping to successfully reintegrate post-prison. Within these confines of extreme social hostility, it remains difficult both to innovate but also to foresee the long-term success of 're-entry' programmes and, to an even greater extent, pre-carceral preventive interventions.

Indeed, social responses of denigration and rejection tend to force coping strategies which appear to entrench isolation and elevate the risk of reoffence (Ricciardelli & Moir, 2013). One such strategy is social withdrawal, where stigmatised individuals limit their social interactions to control situations where others might discover their stigma (Link et al., 2002; Link & Phelan, 2014). In many cases, withdrawal can lead to isolation and the complete avoidance of social contact, entrenching the lack of social support and associated mental health conditions typically experienced by PwSOC (Hannem & Petrunik, 2007; Miner & Munns, 2005). Another identity management approach is generalised self-secrecy, in which individuals conceal their 'whole self' (Perrin, 2018, 2019) in order to bury their pariah status deep below the surface of social life (Winnick & Bodkin, 2009). This construct is acutely present in prison culture, where PwSOC utilise numerous strategies to 'pass' as anything other than a 'sex offender', often declaring themselves to other inmates as having any other conviction across the crime spectrum (Schwaebe, 2005). Other strategies for identity management include 'grouping' and 'denial' (Evans & Cubellis, 2015). In grouping, stigmatised individuals seek out others with similar social status as a means of gaining social support and acceptance. Denial involves the rejection of the socio-legal label attached to PwSOC—an intuitively adaptive and human response (see Blagden et al., 2014; Ware et al., 2015), but one which often carries multiple deleterious implications such as harsher sentencing avenues, barred access to treatment programmes, and more punitive licence (parole) conditions (Lord & Willmot, 2004; Ware et al., 2015; Yates, 2009).

This ecological context for PwSOC, as well as their response to it, is arguably both shaped by and perpetuates a series of mythologies that cyclically reinforce social isolation and reoffending. Furthermore, the mortifications (psychological) and mythologies (social) (perceived as uniquely justified pursuant to widespread myth adoption) that PwSOC endure through social rituals of degradation may further amplify the risk of reoffending (Harper & Bartels, 2018; Quinn et al., 2004). These social processes all commence from heightened affective responses to sexual crimes, and the assimilation of this emotional reasoning into the formulation of a 'sex offender' schema. Myth adoption, then, likely concretises the social construction of the 'sex offender' as an 'irredeemable monster', in a process known as "affect-as-information" (Clore et al., 2001, p.121). As such, representations of this offence type, though often discordant with 'reality' (Harper et al., 2017), in large part set the boundaries of what is possible for future victim prevention initiatives. The compounded effects of social ostracism, internalised stigmatisation, and identity degradation represent structural barriers to re-entry which must be disentangled from policy and practice if reintegration efforts are to be successful. Accordingly, we argue that the uniquely punitive narratives surrounding populist responses to PwSOC, while understandable, constitute major obstacles both in constructive work with this population and for victim prevention by consequence.

Mortification and Degradation

In *The Society of Captives* (1958), Sykes laid out five critical deprivations, or 'pains', of imprisonment which continue to characterise the phenomenological reality of serving time. They are loss of freedom, removal of autonomy, lack of security, severed access to goods and services, and the absence of heterosexual relationships. Scholars have since expanded the concept to include, for example, 'depth' (relating to the feeling of being buried, alive—yet deep below the surface of freedom), 'weight' (referring to the temporal 'heaviness' of seeing out a prison sentence), and 'tightness' (denoting the smothering nature of the total carceral institution) (Crewe, 2011). Living with a pervasive sense of regret about the past, being perpetually worried about the future, and the daily challenge of overcoming inescapable boredom have also been proposed as everyday pains of the prison experience (Irwin, 2006; Johnson & McGunigall-Smith, 2008). While the effects of confinement are far-ranging (see Bukstel & Kilmann, 1980; Liebling & Maruna, 2013), the most severe 'pains' depicted in contemporary prison research tend to be connected to identity (see, inter alia, Crawley & Sparks, 2013; Kerley & Copes, 2009) and "the challenge of cultivating meaning and purpose as time goes by within a Sisyphean regime" (Perrin, 2024, p.32).

Goffman (1961) expanding on this 'pain' at the identity level, argued that this degradation of the self was not only a *feature* but rather a core *function* of any carceral institution. Here, it is argued that a central objective of the prison regime is the stripping of an individual's pre-carceral identity in order to implement a more socially acceptable 'inmate' identity, one which serves to emblematise the ultimate control of the state (Warr, 2020). This is undertaken via a series of 'mortifications', which Goffman (1961) described at length in depicting the inmate's 'total institution'. Upon entering facilities such as prisons, a number of processes are initiated which incrementally dismantle the individual's former identity.

> "*The new inpatient is cleanly stripped of many of his accustomed affirmations, satisfactions, and defenses, and is subjected to a rather full set of mortifying experiences: restriction of free movement, communal living, diffuse authority of a whole echelon of people, and so on. Here one begins to learn about the limited extent to which a conception of oneself can be sustained when the usual supports for it are suddenly removed.*"
>
> (Goffman, 1961, p.148)

Ultimately, the individual is divested of normative social roles and, through ensuing physical and psychological mistreatment, undergoes a self-mortification process. According to Goffman, people serving sentences in total institutions often experience their time in confinement as a deprivation of life in and of itself. Somewhere along this path to existential abyss, the realisation sets in that life beyond the institution will never fully revert to its pre-entrance state. Here, the individual is:

> *"...firmly instructed that the restrictions and deprivations he encounters are not due to such blind forces as tradition or economy – and hence dissociable from self – but are intentional parts of his treatment, part of his need at the time, and therefore an expression of the state that his self has fallen into."*
>
> (Goffman, 1961, p.149)

These processes of prison-based mortification are particularly encompassing for those incarcerated for sexual offence convictions, where a multilayered degradation ceremony looms with exceptional permanence (Blagden et al., 2011; Perrin et al., 2018). Building on Sykes' (1958) framework, individuals in this offence category face not only the standard losses of autonomy, security, and connection but also the additional pains of intense societal disdain and pervasive mistrust from peers and staff. This 'deep burial' beneath layers of ostracism exacerbates the already profound existential weights of confinement, such as those described by Crewe (2011). The compounded effect on identity for those convicted of sexual offences aligns with Goffman's (1961) depiction of mortification as an absolute consequence of the carceral experience. Reintegrating identity-mortified individuals poses a unique challenge for practitioners, who would do well to actively work to deconstruct the prison-based mortification processes discussed above. To this end, effective interventions should focus on identity reconstruction rather than degradation, thus supporting individuals in reclaiming a sense of self that extends beyond their institutionalised identities.

Mythology: Storying the Evil Monster

Fritz Lang's 1931 film 'M' offers an unnerving depiction of a city gripped by fear and moral panic, centring on a child murderer who terrorises Berlin. The police and a vigilante group compete in their desperate hunt for the killer, with the leader of the latter proclaiming "This monster has no right to live. He must disappear. He must be exterminated, without pity … without scruples!". The film hints at a sexual motivation behind the crimes, although this is never explicitly stated—the mere suggestion is enough to diffuse an atmosphere of disgust. The film implicitly places the 'sexual offender' in a unique category of depravity, subtly drawing a distinction between child murder and sexual offence, the latter representing an especially unspeakable evil. Of course, the film's portrayal reflects the broader tendency of society to react to sexual crimes with an exceptional revulsion—one that often overrides any rational discourse and willingness to explore solutions beyond retributive reactivity.

This reluctance to engage with the complexities of sexual crime has remained a constant in the Hollywoodisation of offending behaviour, where 'sexual offenders' are routinely cast as irredeemable monsters. In The Silence of the Lambs (Demme, 1991), Hannibal Lecter and Buffalo Bill are portrayed as terrifying, psychosexual predators, cementing a cultural image of the sexually motivated criminal as something other than human—there are even efforts to present these characters as 'superhuman' (hyperintelligent, masterfully calculating, physically superior etc.), which contributes to the notion of such individuals being limitlessly 'unreachable' by society's law-abiding average Joe. Similarly, A Clockwork Orange (Kubrick, 1971) features Alex, a violent man whose sexual assaults are depicted as part of his inherently evil predisposition for depravity, reinforcing the association between sexual violence and moral corruption. In The Green Mile (Darabont, 1999), John Coffey is falsely accused and convicted of sexually assaulting and murdering two young girls, a crime that propels the narrative into a reflection on the moral decay of society.

The public's response to Coffey, again marked by vociferous repulsion, underscores the heightened fear and horror reserved for sexual crimes, particularly those involving children. Coffey's daunting physique and initial portrayal as quietly menacing catalyse the public's eagerness to see him as the embodiment of evil. Even as the film unfolds, and Coffey is revealed as benevolent, saintly, and (most crucially) innocent, the narrative soon refocuses on storying another archetypal monster—the true perpetrator—in the same, almost formulaic way. This storying, or—monsterisation—mirrors the real-world tendency to respond to sexual crime with a reserve of revulsion that surpasses other forms of violence. These portrayals fuel public fears and entrench myths about the nature of sexual offending, making it difficult to discuss the issue in a way that fosters understanding and prevention. The cinematic monsterisation of PwSOC mirrors society's deep discomfort with sexual crime, resulting in punitive, reactionary measures that focus on punishment rather than prevention:

> Indeed, the public endorse numerous incorrect beliefs (or myths) about sexual offenders that are thought to be shaped by the media. These include believing that sexual offenders are homogeneous, primarily strangers, predatory, re-offenders, specialists, resistant to treatment, products of early abuse, mentally ill, and/or 'dirty old men' (Harper & Bartels, 2018, p.277)

Of course, societal responses to sexual offences arise from a complex interplay of factors, including the public's desire for safety, political pressure to address these concerns with straightforward legislation, sensationalised media coverage, exaggerated reoffence statistics, and heightened parental fears in an increasingly uncertain world. Consequently, public perception of sexual offence recidivism is generally inflated. For instance, participants in research from Levenson et al. (2007) estimated that, on average, 76% of child molesters reoffend. Another study by Scurich et al. (2016) revealed participant estimates of 74% recidivism for rapists. However, empirical research on sexual offence recidivism suggests much lower rates. Studies defining recidivism as new legal charges or convictions for sexual crimes generally find rates of 5–15% after 5 years and 10–25% after 10+ years.

One of the most widely cited meta-analyses of 61 studies found a mean sexual recidivism rate of 13.4% within four to five years

(Hanson & Bussiere, 1998), with almost identical results at follow-up (Hanson & Morton-Bourgon, 2005). Other studies have found even lower rates, such as 3.5% within 3 years (Langan et al., 2003; Cohen, 2018). Certainly, it is important to consider that the true rate of sexual recidivism is likely to be higher than these projections due to underreporting, as per the "dark figure of crime" (Calkins et al., 2014). Even so, factoring in life course perspectives and the age-crime curve paradigm within desistance research, it would appear that "the pessimistic notion of the irredeemable criminal simply does not fit one of the best-established empirical findings in criminology: Almost everyone who is labelled as a super predator eventually "goes straight"—or desists—from crime" (Maruna, 2001, p. 6). Nevertheless, a number of engrained responses to sexual crime continue to shape the options available for community-based prevention efforts.

The widespread tendency to use 'capture all' labels to describe 'sex offenders' as one homogeneous group of predators serves as another central justification for society's extreme response. High-profile cases, such as stranger kidnappings and murders of young children, shape the public's understanding of sexual crimes. This is of course fuelled by political discourses and media outlets that exploit affect heuristic cognitive processes that conflate severity with frequency (Harper & Hogue, 2017). News media sources, political figures, and even some academic papers use the terms 'sex offender' and 'sexual predator' interchangeably. As a result, although the 'predator' label—typically associated with crimes like serial rape, sexually motivated murder, and child kidnapping (Quinn et al., 2004)—applies to only a small subset of PwSOC (Ministry of Justice, 2013), all PwSOC are commonly cast as dangerous psychopaths, regardless of their specific offence classification.

Repercussions for Reintegration

Identity mortification is particularly salient for PwSOC, who are subjected to an enduring societal stigma that is not only reinforced by their legal label but also amplified by monsterising mythologies. These mythological portrayals make ideals of 'rehabilitation' appear not only unlikely, but naïve, 'soft', and even anti-social. This widespread cultural reinforcement of fear feeds a societal 'unreadiness' for change. It also inhibits discourse about prevention, rehabilitation, and reintegration. As a result, interventions that might otherwise serve as preventive measures face resistance, both in policy and in practice.

To further substantiate this argument, ample research has shown how media portrayal shapes public perceptions of crime (see, inter alia, Näsi et al., 2021; Sherrill, 2022; Zillmann & Wakshlag, 2013). Jewkes (2004) argues that the mass media not only reflects but constructs social reality, particularly in the context of crime and 'deviance'. Additionally, Harper and Hogue (2017) highlight how distorted portrayals of PwSOC in the media exacerbate public fear and perpetuate myths, such as the idea that 'sexual offenders' are a homogenous group who cannot be rehabilitated. These empirical works reinforce the need to challenge and critically examine the role of media in shaping public understanding of sexual offences. People appear bonded in their proneness to effortlessly form attitudes on the basis of "media exposure, and other forms of socially supplied information" (Crano et al., 2010, p.3). Accordingly, media coverage about sexual offenders is likely to play a major role in shaping the public's attitude toward sexual offenders (Harper & Hogue, 2015; Malinen et al., 2014; Quinn, Forsyth, & Mullen-Quinn, 2004; Thakker, 2012).

Despite this relatively bleak picture, there has been an emergence of some hopeful community-based prevention programmes. Circles of Support and Accountability (CoSA)—a community-based reintegration programme founded on peer helping principles and civic engagement—yielded an 83% reduction in sexual recidivism, a 73% reduction in all types of violent recidivism, and an overall reduction of 71% in all types of recidivism in comparison to matched offenders (Wilson et al., 2009). CoSA involves trained volunteers forming a support circle around a 'Core Member', providing practical assistance, emotional guidance, and accountability. Through regular meetings and a restorative justice approach, CoSA encourages personal growth and positive transformation, promoting a holistic alternative to traditional criminal justice approaches (Lowe & Willis, 2020). While these initiatives are not extensively researched at present, they point to an emerging theme, which is that meaningful change happens in the milieu of reciprocity, community, and connectedness.

CoSA's efficacy appears rooted in its collaborative approach to the problem of crime, viewing it not as an isolated malevolent pursuit by individuals but rather a result of systemic and relational issues. It is therefore one of a constellation of innovative initiatives that connect to an evolving generative movement within the justice space. This new direction of 'generative justice' (McNeill et al., 2023) marks a decisive movement from the deficit-based, 'corrective', and retributive models that have historically prevailed to a strengths-based approach which prioritises the restoration of social relations, community solidarity, and inclusion. Generative justice recognises that formerly crime-involved individuals possess vital insights into the desistance process and effective strategies to facilitate it. Crucially, however, it simultaneously acknowledges that desistance does not occur in social vacuums and should be collaborative. CoSA is emblematic of generative justice and therefore affords new possibilities for both responses to and desistance from crime. However, such preventive initiatives can only ever be as effective as their receiving social climate deems them to be.

The belief that sex offenders are unredeemable predators is so widely accepted in modern society that it provides a rallying point for victims' groups, politicians, and others with the desire to build consensus on the need to make penal sanctions harsher, which tends to dovetail with an abandonment of community-based prevention initiatives.

All of this reinforces dualistic thinking, with the socially 'safer' side being the more punitive space. The end result appears to denote a choice between belonging to dichotomous groupings such as good or evil, nature or nurture, vigilantism or due process, retribution, or restoration et cetera. With vengeful narratives prevailing, at least in the ecological niche of sexual offence, it remains difficult to imagine how an individual can go about 'making good' and rebuilding their lives following a sexual offence conviction. As one such individual, who denied his offence for many years, explained in research from Blagden et al. (2011):

> "*What's the worst thing you can say to someone? "I'm Hitler". It's not self-preservation for you, it's the words coming out of your mouth, they can't, the*

> *fact that that's what you did, it's not, I don't know, the whole character you've been building up for 39 years ... it's like saying I'm Hitler ... why would you own up to that? ... it would have been easier if I murdered her, I don't mean that ... err ... but on the outside you'd get less stigma."*
>
> Martin (Blagden et al., 2011, p. 568)

The notion of 'being' a 'murderer' having a less deleterious impact on one's identity than 'being' a 'sex offender' is well established in phenomenological work. As the partner of one convicted person in research from Blagden et al. (2011) asserted: "I'd rather you had committed any offence other than what you've done, even murder, I could've accepted that." The internal and external shaming faced by PwSOC present significant barriers for desistance at the individual level, but also obstruct preventive efforts at the social and structural level. These layers of stigmatisation, reciprocally enforced by processes of mortification and mythology, not only confound rehabilitation processes but also restrict opportunities for community-based initiatives that could address sexual offence root causes without attempting to 'excuse' offending behaviour.

Concluding Comments

This chapter has shown that mortification and mythology, as entwined social processes, shape both the experiences of individuals with sexual offence convictions and the societal responses existing around them in engrained ways. Readers may judge the value of examining these heuristic devices as constituting specific challenges that influence the criminal justice system and community-based responses to PwSOC. In the midst of politically charged discourse on 'sex offenders', the chapter highlights the importance of acknowledging the psychological impact of these processes on individuals, especially regarding their reintegration into communities. At the same time, it calls for a careful separation of these personal experiences from the mythologies that intensify public reaction to this category of offence. Though it may seem unscholarly, even crass, to invoke 'Hollywoodised' portrayals of sexual offences to make sense of reintegration barriers, it is evident that entertainment media—particularly via movie portrayal—has the capability of permeating human cognition in ways that may influence critical thinking and the knowledge construction of social phenomena. The schematic frameworks people use to understand complex social constructs are frequently shaped by artistic portrayals, often in simplified or dichotomous forms that can seem harmless or even benevolent. These influences are formative, often laying the groundwork for ideologically driven thinking at more nuanced levels. In the case of sexual offences, media-driven distortions reinforce accepted, yet often inaccurate, beliefs about both the crimes and the individuals who commit them. If society is to take the collective responsibility of addressing sexual abuse seriously, these entrenched misconceptions must be disentangled from discourse surrounding policy and practice.

With the help of mortification and mythology, society has innovated pervading methods of expulsion for those with sexual offence criminal convictions. However, successful reintegration can only ever be "a two-way process, requiring both effort on the part of the former prisoner (e.g., desistance, repentance), but also on the part of some wider community (e.g., forgiveness, acceptance)" (Maruna, 2011, p.13). Degradation has no place in a restorative and, most importantly, preventive justice approach which values whole communities. The identity transformation required for societal re-entry must be a collaborative endeavour, requiring symbolically affirming 'rituals' which recognise efforts for redemption and respond with a genuine commitment to societal forgiveness and inclusion. Intervening to prevent the deleteriousness of mortification, while working to actively challenge the mythology enveloping sexual offences is essential for fostering societal change. This must be done not only through policy reform but also through grassroots efforts that prioritise collaborative, community-oriented, and solidaristic approaches to justice (McNeill et al., 2023). Public perceptions shaped by media need to be actively challenged through educational interventions and a greater focus on community participation. Reformation of social policy within this arena, combined with a shift in societal attitudes, is crucial to breaking cycles of social ostracism and reoffending, while envisioning a more preventive approach to justice.

References

Blagden, N. J., Winder, B., Thorne, K., & Gregson, M. (2011). 'No-one in the world would ever wanna speak to me again': An interpretative phenomenological analysis into convicted sexual offenders' accounts and experiences of maintaining and leaving denial. *Psychology, Crime & Law, 17*(7), 563–585.

Blagden, N., Winder, B., Gregson, M., & Thorne, K. (2013). Working with denial in convicted sexual offenders: A qualitative analysis of treatment professionals' views and experiences and their implications for practice. *International Journal of Offender Therapy and Comparative Criminology, 57*(3), 332–356.

Blagden, N., Winder, B., Gregson, M., & Thorne, K. (2014). Making sense of denial in sexual offenders: A qualitative phenomenological and repertory grid analysis. *Journal of Interpersonal Violence, 29*(9), 1698–1731.

Bukstel, L. H., & Kilmann, P. R. (1980). Psychological effects of imprisonment on confined individuals. *Psychological Bulletin, 88*(2), 469.

Calkins, C., Jeglic, E., Beattey, R. A. Jr., Zeidman, S., & Perillo, A. D. (2014). Sexual violence legislation: A review of case law and empirical research. *Psychology, Public Policy, and Law, 20*(4), 443–462.

Clore, G. L., Gasper, K., & Garvin, E. (2001). Affect as information. In J. P. Forgas (Ed.), *Handbook of affect and social cognition* (pp. 121–144). Lawrence Erlbaum.

Cohen, T. H. (2018). Predicting sex offender recidivism: Using the Federal Post Conviction Risk Assessment Instrument to assess the likelihood of recidivism among federal sex offenders. *Journal of Empirical Legal Studies, 15*(3), 456–481.

Crano, W. D., Cooper, J., & Forgas, J. P. (2010). Attitudes and attitude change: An introductory review. In J. P. Forgas, J. Cooper, & W. D. Crano (Eds.), *The psychology of attitudes and attitude change* (pp. 3–18). Psychology Press.

Crawley, E., & Sparks, R. (2013). Older men in prison: Survival, coping and identity. In A. Liebling & S. Maruna (Eds.), *The effects of imprisonment* (pp. 343–365). Willan.

Crewe, B. (2011). Depth, weight, tightness: Revisiting the pains of imprisonment. *Punishment & Society, 13*(5), 509–529.
Darabont, F. (Director). (1999). *The green mile* [Film]. Warner Bros.
Demme, J. (Director). (1991). *The silence of the lambs* [Film]. Orion Pictures.
Evans, D. N., & Cubellis, M. A. (2015). Coping with stigma: How registered sex offenders manage their public identities. *American Journal of Criminal Justice, 40*, 593–619.
Fortney, T., Levenson, J., Brannon, Y., & Baker, J. N. (2007). Myths and facts about sexual offenders: Implications for treatment and public policy. *Sexual Offender Treatment, 2*(1), 1–17.
Gavin, H. (2005). The social construction of the child sex offender explored by narrative. *The Qualitative Report, 10*(3), 395–413.
Goffman, E. (1961). *Asylums: Essays on the social situation of mental patients and other inmates.* Doubleday Anchor Books.
Hannem, S., & Petrunik, M. (2007). Circles of support and accountability: A community justice initiative for the reintegration of high-risk sex offenders. *Contemporary Justice Review, 10*(2), 153–171.
Hanson, R. K., & Bussiere, M. T. (1998). Predicting relapse: A meta-analysis of sexual offender recidivism studies. *Journal of Consulting and Clinical Psychology, 66* (2), 348–362.
Hanson, R. K., & Morton-Bourgon, K. E. (2005). The characteristics of persistent sexual offenders: A meta-analysis of recidivism studies. *Journal of Consulting and Clinical Psychology, 73*(6), 1154–1163.
Harkins, G. (2020). *Virtual pedophilia: Sex offender profiling and US security culture.* Duke University Press.
Harper, C. A., & Bartels, R. M. (2018). Implicit theories and offender representativeness in judgments about sexual crime. *Sexual Abuse, 30*(3), 276–295.
Harper, C. A., & Hogue, T. E. (2015). The emotional representation of sexual crime in the national British press. *Journal of Language and Social Psychology, 34*(1), 3–24.
Harper, C. A., & Hogue, T. E. (2017). Press coverage as a heuristic guide for social decision-making about sexual offenders. *Psychology, Crime & Law, 23*(2), 118–134.
Irwin, J. (2006). *The warehouse prison: Disposal of the new dangerous class.* Oxford University Press.
Jewkes, Y. (2004). *Media and crime.* Sage.

Johnson, R., & McGunigall-Smith, S. (2008). Life without parole, America's other death penalty: Notes on life under sentence of death by incarceration. *The Prison Journal, 88*(2), 328–346.
Kerley, K. R., & Copes, H. (2009). Keepin'my mind right identity maintenance and religious social support in the prison context. *International Journal of Offender Therapy and Comparative Criminology, 53*(2), 228–244.
Kubrick, S. (Director). (1971). *A clockwork orange* [Film]. Warner Bros.
Langan, P. A., Schmitt, E. L., & Durose, M. R. (2003). Recidivism of sex offenders released from prison in 1994. US Department of Justice. https://www.bjs.gov/content/pub/pdf/rsorp94.pdf
Levenson, J. S., Brannon, Y., Fortney, T., & Baker, J. (2007). Public perceptions about sex offenders and community protection policies. *Analyses of Social Issues and Public Policy, 7*(1), 1–25.
Liebling, A., & Maruna, S. (Eds.). (2013). *The effects of imprisonment.* Routledge.
Link, B. G., & Phelan, J. (2014). *Stigma power. Social science & medicine, 103,* 24–32.
Link, B. G., Struening, E. L., Neese-Todd, S., Asmussen, S., & Phelan, J. C. (2002). On describing and seeking to change the experience of stigma. *Psychiatric Rehabilitation Skills, 6*(2), 201–231.
Lord, A., & Willmot, P. (2004). The process of overcoming denial in sexual offenders. *Journal of Sexual Aggression, 10*(1), 51–61.
Lowe, G., & Willis, G. (2020). "Sex offender" versus "person": The influence of labels on willingness to volunteer with people who have sexually abused. *Sexual Abuse, 32*(5), 591–613.
Malinen, S., Willis, G. M., & Johnston, L. (2014). Might informative media reporting of sexual offending influence community members' attitudes towards sex offenders? *Psychology, Crime & Law, 20*(6), 535–552.
Maruna, S. (2001).*Making Good: How Ex-offenders. Reform and Reclaim their Lives.* APA
Maruna, S. (2011). Reentry as a rite of passage. *Punishment & Society, 13*(1), 3–28.
McNeill, F., Corcoran, M., & Weaver, B. (2023). Generative justice: Exploring how communities re/create solidarity after crime and punishment. Retrieved Aug 22, 2025, from https://www.isrf.org/fellows-projects/generativejustice/%23%3A%7E%3Atext%3DBy%20'Generative%20Justice'%2C%20we,relations%20characterised%20by%20increased%20solidarity
Miner, M. H., & Munns, R. (2005). Isolation and normlessness: Attitudinal comparisons of adolescent sex offenders, juvenile offenders, and

nondelinquents. *International Journal of Offender Therapy and Comparative Criminology, 49*(5), 491–504.

Ministry of Justice, Home Office, & Office for National Statistics. (2013). *An overview of sexual offending in England and Wales (Statistics Bulletin).* https://assets.publishing.service.gov.uk/government/uploads/system/uploads/attachment_data/file/214970/sexual-offending-overview-jan-2013.pdf

Näsi, M., Tanskanen, M., Kivivuori, J., Haara, P., & Reunanen, E. (2021). Crime news consumption and fear of violence: The role of traditional media, social media, and alternative information sources. *Crime & Delinquency, 67*(4), 574–600.

Perrin, C. (2018). Coping with incarceration: The emerging case for the utility of peer support programs in prisons. In A. Mills. & K. Kendall (Eds.), *Mental health in prisons: Critical perspectives on treatment and confinement* (pp. 331–351). Palgrave Macmillan.

Perrin, C. (2019). Peer support and individuals with sexual convictions: Complementing traditional rehabilitation strategies. In N. Blagden, B. Winder, K. Hocken, R. Lievesley, P. Banyard, & H. Elliott (Eds.), *Sexual crime and the experience of imprisonment* (pp. 224–244). Palgrave Macmillan.

Perrin, C. (2024). *Peer support in prison: How incarcerated people make meaning through active citizenship.* Routledge.

Perrin, C., Blagden, N., Winder, B., & Dillon, G. (2018). It's sort of reaffirmed to me that I'm not a monster, I'm not a terrible person: Sex offenders' movements toward desistance via peer-support roles in prison. *Sexual Abuse, 30*(7), 759–780.

Quinn, J. F., Forsyth, C. J., & Mullen-Quinn, C. (2004). Societal reaction to sex offenders: A review of the origins and results of the myths surrounding their crimes and treatment amenability. *Deviant Behavior, 25*(3), 215–232.

Ricciardelli, R., & Moir, M. (2013). Stigmatised among the stigmatised: Sex offenders in Canadian penitentiaries. *Canadian Journal of Criminology and Criminal Justice, 55*(3), 353–386.

Schwaebe, C. (2005). Learning to pass: Sex offenders' strategies for establishing a viable identity in the prison general population. *International Journal of Offender Therapy and Comparative Criminology, 49*(6), 614–625.

Scurich, N., Gongola, J., & Krauss, D. A. (2016). The biasing effect of the "sexually violent predator" label on legal decisions. *International Journal of Law and Psychiatry, 47*, 109–114.

Sherrill, L. A. (2022). The serial effect and the true crime podcast ecosystem. *Journalism Practice, 16*(7), 1473–1494.

Thakker, J. (2012). Public attitudes to sex offenders in New Zealand. *Journal of Sexual Aggression, 18*(2), 149–163.

Ware, J., Marshall, W. L., & Marshall, L. E. (2015). Categorical denial in convicted sex offenders: The concept, its meaning, and its implication for risk and treatment. *Aggression and Violent Behavior, 25*, 215–226.

Warr, J. (2020). "Always gotta be two mans": Lifers, risk, rehabilitation, and narrative labour. *Punishment & Society, 22*(1), 28–47.

Wilson, R. J., Cortoni, F., & McWhinnie, A. J. (2009). Circles of support & accountability: A Canadian national replication of outcome findings. *Sexual Abuse, 21*(4), 412–430.

Winnick, T. A., & Bodkin, M. (2009). Stigma, secrecy, and race: An empirical examination of black and white incarcerated men. *American Journal of Criminal Justice, 34*, 131–150.

Yates, P. M. (2009). Is sexual offender denial related to sex offence risk and recidivism? A review and treatment implications. *Psychology, Crime & Law, 15*(2–3), 183–199.

Yung, C. R. (2011). Sex offender exceptionalism and preventive detention. *J. Crim. L. & Criminology, 101*, 969.

Zillmann, D., & Wakshlag, J. (2013). Fear of victimization and the appeal of crime drama. In *Selective exposure to communication* (pp. 141–154). Routledge.

2

Examining the Tensions Between Policing People Convicted of Sexual Offending and Promoting the Desistance Process

Stephanie Kewley and Sarah Pemberton

Introduction

Professionals working with individuals convicted of sexual offending are committed to ensuring that further sexual offending does not occur. Underpinning Western criminal justice approaches to the prevention of sexual reoffending is the notion that people who commit crime must be both punished and controlled, while, at the same time, helped through reintegrative strategies to desist from further criminal activity. However, control and containment policies, while designed to protect the public, are likely to also result in unintended consequences for those convicted of sexual offending. Such consequences result in reinforced stigma, unemployment, social isolation, and insecure housing, all of which hinder the desistance process; and may inadvertently place people at risk of further

S. Kewley (✉)
Liverpool John Moores University, Liverpool, UK
e-mail: s.cain@ljmu.ac.uk

S. Pemberton
Birmingham City Univeristy, Birmingham, UK

N. Blagden et al. (eds.), *Sexual Crime and Community Reintegration*, Sexual Crime,
https://doi.org/10.1007/978-3-032-06034-1_2

offending. Thus, formal criminal justice agents, such as specialist police offender managers, practice within a moral tension. Not only are they required to risk assess and risk manage people with sexual convictions, but they must do so while helping them navigate a process of desistance in a highly politicised climate of public protection. This chapter asks how formal criminal justice structures, specifically the police, can assist the process of desistance for those convicted of sexual offences amidst these tensions. To do this, we first examine the role of the police within the Multi-Agency Public Protection Arrangements (MAPPA) in preventing sexual reoffending. We then provide a brief overview of desistance before considering the work of specialist police teams in the Management of Sexual or Violent Offenders detailing the moral tensions they face, with suggestions on how this could be navigated.

The Role of the Police Within Mappa

MAPPA aims to protect the public and prevent further serious harm from known 'offenders.' They were established under the Criminal Justice and Court Service Act (2000) and the Criminal Justice Act (2003) across England and Wales. The legislation mandates police, probation, and prison services with 'responsible authority' to local MAPPA boards and other agencies (such as health, education, and employment services) with a 'duty to co-operate.' Relevant agencies then assess and manage MAPPA-eligible offenders by working together, sharing information, and implementing management plans (*MAPPA Guidance*, 2021). Following sentencing, people are allocated into one of three MAPPA categories:

- Category 1 includes all those required to comply with notification requirements as set out in the Sexual Offences Act (2003); this group is commonly referred to as Registered Sexual Offenders (RSOs)
- Category 2 includes those convicted of murder or other violent/sexual offending (but not requiring notification as in Category 1)

- Category 3 includes people who do not meet the criteria for Category 1 or 2 but have committed an offence that indicates they present a risk of causing serious harm and, thus, require multi-agency level management.

MAPPA management arrangements are further designed to ensure those with the greatest level of perceived risk are provided with the greatest level of multi-agency cooperation and are, thus, further organised by three levels of management:

- Level 1 indicates a person's risk level can be managed by a single or lead agency; that is, the lead agency is able to implement a risk management plan that addresses and manages identified risks without the need for multi-agency meetings.
- Level 2 cases are assessed as presenting a high or very high risk of serious harm; such cases require multi-agency involvement in the delivery of interventions or services to address the presenting risks. Level 2 cases are managed through formal multi-agency meetings.
- Level 3 cases meet the criteria of Level 2. However, senior management representation from 'responsible authority' and 'duty-to-co-operate' agencies is needed so that resources can be committed to manage the high or very high risk of serious harm. Level 3 cases might also be high-profile cases in which significant media or public attention is present.

The volume of cases managed by MAPPA has continued to increase annually, with a current caseload of 87,657 (MAPPA, 2021). Much of this caseload (98.4%) is managed at MAPPA Level 1, of which 73% (63,827) are Category 1 RSOs. In the most recent annual report (March 2021), 2% (1,641) of RSOs were cautioned or convicted for a breach of notification requirement (not a further sexual offence), and of those with eligible license conditions (1,361), 18% were returned to custody following a breach of license (also not a further sexual offence). Within this period, 43 RSOs were, however, convicted of a Serious Further Offence (SFO). Although even one further sexual offence is unacceptable, this number represents only 0.07% of the total RSO population.

Given such low reoffending rates, it would appear that the MAPPA approach is effective in preventing sexual reoffending. However, testing this assumption is challenging, as MAPPA-eligible individuals automatically commence management through MAPPA after their sentence, which makes it extremely difficult to undertake a rigorous control group analysis (Kemshall & Hilder, 2020). Further, this success rate may not be solely attributed to MAPPA's approach; sexual recidivism rates are low across the board. Data from other countries that do not use MAPPA, as well as pre-MAPPA records, show that within 10 to 15 years of living offence-free in the community, the large majority of those convicted of a sexual crime have no greater risk of sexual offending than people who have never been convicted of a sexual offence at all (Hanson *et al.*, 2018).

Sexual offending is a complex crime, so caution is needed when considering these findings; not all sexual assaults (including re-offences) are reported or result in a conviction. As such, these figures might not represent an accurate scale of sexual recidivism. Despite such low 'known' reoffending rates, those convicted of a sexual offence remain under MAPPA arrangements, and those living in the community are managed by either the probation service, the police, or both. For the purpose of this chapter, attention focuses on the work of the police[1] in England and Wales. Little is known of the effectiveness of specialist police teams responsible for the management of people with sexual convictions (Thomas, 2016); therefore, this is an area requiring further scrutiny.

Police unit structures vary across each force, with RSOs managed by specialist teams of offender managers generally located within Public Protection Units (College of Policing, 2020b) and tasked with the Management of Sexual or Violent Offenders (MOSOVO). MOSOVO teams are trained, experienced police officers and staff who identify, risk assess, and manage MAPPA cases. While operating in a risk-based and highly politicised model of public protection (McCartan & Gotch,

[1] For a detailed discussion regarding public protection by probation services in England and Wales, please see Kewley and Brereton (forthcoming)

2020), risk management strategies incorporate the approach developed by Kemshall and colleagues at De Montfort University known as the "4 Pillars of Risk Management" (Public Protection Group, 2020, p. 31). The 4 Pillars require offender managers to incorporate aspects of Supervision; Monitoring and Control; Interventions and Treatment; and Victim Safety Planning into all risk management plans. It is claimed that this blend of control strategies and reintegrative approaches can provide a criminal justice model that punishes *and* rehabilitates (Kemshall & Hilder, 2020; Maruna & Mann, 2019). Nevertheless, scholars note that significant barriers are experienced by people attempting to desist, as their reality is a context dominated by pervasive punishment and controls (McNeill, 2019). Indeed, for people convicted of sexual offending, this reality is even harsher, with containment, control, and preventative sentencing dominating criminal justice policy and practice (Kewley & Brereton, forthcoming). Thus, given the critical public protection role MOSOVO offender managers play when working with people with sexual convictions, consideration is needed to understand if desistance can be promoted. However, before we consider this, we briefly introduce the notion of desistance.

Desistance

Desistance is a process by which people cease criminal behaviour; there is no single moment at which people become desisters, nor is progress evenly paced, linear, or always in a forward direction. Instead, people zigzag into desistance (Maruna, 2001), with periods of intermittency, indecisiveness, ambivalence, lapse, and even relapse (Ouellet, 2019; Piquero *et al.*, 2013). There are two distinguishable phases to successful desistance in which a blend of internal and social shifts is required (LeBel *et al.*, 2008). The primary phase sees crime cease; this is purely behavioural and requires a secondary cognitive phase in which the person shifts their identity from 'offender' to 'non-offender.' This secondary phase is two-fold, requiring an internal psychological transformation in which the person disassociates with the 'offender' label, and external validation from others (Farrall *et al.*, 2010; Giordano *et al.*, 2002; Maruna,

2001). Both internal and external validation assists with the process of de-labelling, as both the desister and those interacting with them no longer perceive them as an 'offender'; thus, the person regains a sense of belonging to their community (McNeill, 2016). Therefore, for people to 'go straight,' cognitive and behavioural changes made at the individual level are only strengthened, and realised by and with, the social structures and networks in which they exist (Bronfenbrenner, 1977; Weaver, 2016).

Theories of desistance can broadly be categorised into those that explain the process through individual/agentic, social/structural, interactional, and situational theories. Weaver (2019) provides an extensive critique of these; therefore, for the purpose of this chapter, we provide a very brief summary. Individual or agentic theories draw on ontogenetic explanations, that is, those examining the relationship between age/maturation and crime (Glueck & Glueck, 1950; Laub & Sampson, 2001). These theories describe how the onset and persistence of crime usually occurs in early adolescence, peaking in late adolescence and often declining into adulthood (late 20s/30s) (Hirschi & Gottfredson, 1983). Agentic explanations, or cognitive processes, are explained through rational choice theories (Clarke & Cornish, 1985; Cornish & Clarke, 1987; Paternoster & Bushway, 2009), which suggest people make decisions to offend (or not) based on cost/benefit type calculations. For example, if ceasing offending is deemed to provide greater benefits than continuing offending, then the person will desist. Alternatively, social and structural theories are those that involve factors external to the actor (sociogenic). These theories highlight the importance of informal ties or social bonds with institutions such as family, education, and employment (Sampson & Laub, 2003). Where informal ties are weak or broken (particularly during adolescence), desistance from crime becomes a greater challenge (Hirschi, 1969). Pro-social bonds or associations increase exposure to pro-social capital and, by default, provide greater opportunities for education, employment, and intimate relationships with non-offending peers (Matza, 1964).

Interactionist theorists argue that desistance occurs through a delicate interplay of agency and structure, as both are strongly interconnected. Desistance occurs as a result of either a changed socio-structural context in which cognitive and behavioural changes (and ultimately identity transformation) occur or vice versa (Farrall *et al.*, 2010; Giordano *et al.*, 2002; Healy, 2013; Maruna, 2001). The degree of weight or sequencing is, of course, very much debated (LeBel *et al.*, 2008). Finally, situational theories of desistance are used to explain the role of time, space, and place in the process (Flynn, 2017). Environmental context plays a key role in both the persistence and desistance of crime (Farrall *et al.*, 2014). Unsurprisingly, those who desist from crime tend to engage in daily routines and activities in spaces and places where work, education, and family life co-exist. Thus, an important consideration in understanding the process of desistance is the spaces and places in which people inhabit.

Facilitating the mechanisms that help promote desistance and support those convicted of sexual crimes is crucial as they help a) prevent further harm to victims, b) reduce the social and economic cost to society, and c) help people safely contribute to society. However, a comprehensive understanding of *who* and *how* formal criminal justice structures can 'assist desistance' is limited (Cresswell, 2020; Villeneuve *et al.*, 2021). Knowledge is particularly sparse in populations of people convicted of sexual offending. Thus, we now turn to the tensions that exist between MOSOVO policy and practice, and the effort recognised to promote desistance effectively. In particular, we focus on the tensions that exist in managing this group due to legislative barriers and the pervading negative attitudes towards people with sexual convictions.

Can MOSOVOs Assist Desistance?

An authentic sense of agency and self-determination (coupled with external opportunities) provides the bedrock to realise identity transformation, behaviour change, and desistance (Giordano *et al.*, 2008; King, 2014). This can sometimes be realised *despite* external controls and sanctions; however, pervasive criminal justice controls (McNeill, 2019) and punitive legal restrictions required by people with sexual convictions (Thomas & Marshall, 2021) will likely only hinder this process. One way to potentially counter this is for MOSOVOs to support the development of a new non-offending identity, reduce stigma and foster social capital.

It is well documented that society holds hostile views toward people who have committed sexual offences (McCartan *et al.*, 2015), which often results in feelings of shame and stigma (Bailey & Klein, 2018). A particular challenge for any professional supporting a person with a sexual conviction, therefore, is to recognise not only the difficulties this unsympathetic context brings but also the impact internalised shame can have on successful transformation into desistance. One particular concern is that the reinforcement of stigma not only has detrimental consequences on risk, but may also prevent people from seeking help, making it harder to help them desist from crime (Grady *et al.*, 2019). When shame is stigmatised, the person is degraded, labelled (in this case, 'sex offender' or 'deviant'), and excluded, making cognitive transformation a much greater challenge. However, with reintegrative shame, the individual still experiences feelings of remorse and guilt, but their sense of worth is preserved and not labelled, and desistance is more likely (Braithwaite, 1989). Derived from labelling theory, the interplay between one's own identity (self-label) interacts with the perception or label of others as a result of social interaction. True desistance is experienced when the new non-offending identity is both internalised and fully accepted and recognised by others (either informally and/or formally) (Buchanan & Krohn, 2020).

Interactions with formal agents, such as MOSOVOs, can, therefore, both reduce and reinforce stigma. While MOSOVO-specific studies are lacking in the extant literature, general research finds a mix of experiences from people convicted of sexual offending. Individuals have reported feeling supported by their probation officer (Mann et al., 2018), and they felt their supervisor expressed care and concern, and had a personal interest in them (Farmer *et al.*, 2015). Individuals also felt that formal agents believed they could change (Blagden *et al.*, 2016) and that they felt safe to discuss sensitive matters without feeling judged (Winder *et al.*, 2020). Such experiences likely reflect positive relationships in which the formal agent promotes the desistance process by reducing stigma, enabling people to develop new identities that depict a positive future self (Paternoster & Bushway, 2009) and providing support to re-story new narratives (Mullins & Kirkwood, 2019) in ways that are meaningful and realistic. Of the few MOSOVO studies undertaken, it was found that MOSOVOs tend to perceive the provision of welfare and support as a detraction from their core proactive policing duties (Christensen *et al.*, 2021; Kewley, 2017; Nash, 2019). MOSOVOs view people with sexual convictions as groups who should be closely managed because they are dangerous, untrustworthy, manipulative, and need to be caught out (Kewley, 2017), they are viewed as monsters (Nash, 2016), and if given the opportunity, their risk would escalate (Mann et al., 2018). Such attitudes will likely permeate interactions between the MOSOVO and the people they supervise. Indeed, despite some of Mann et al.'s (2019) participants reporting positive experiences, others report feelings of prejudice, not feeling trusted, or feeling judged (Kras, 2019), experiencing hostile supervisory tactics that created resistance and believing their crimes repel staff (Farmer *et al.*, 2015). These findings are perhaps unsurprising as MOSOVOs operate within models of public protection and control and containment frameworks. These models do little to promote agency or identity transformation and, as such, are incongruous with the factors needed to promote the desistance process (Mann *et al.*, 2019).

To effectively assist desistance, formal agents must promote hope and optimism and convey a belief that the person attempting to desist can change (McAlinden *et al.*, 2016). However, formal agents' attitudes are often incompatible with this objective; MOSOVOs tend not to perceive this group in a favourable light and, as a result, design risk management plans dominated by strategies of control (Kewley *et al.*, 2020). Again, this is understandable as MOSOVOs, first and foremost, work within a framework of public protection (McCartan & Gotch, 2020) that enforces court-ordered conditions using surveillance and risk management, and at the core of their work requires proactive policing (Mann & Lundrigan, 2021). Indeed, being enforcers of prospective sentences, and engaging in regular risk assessment and risk management activities, will inevitably reinforce attitudes and behaviours that are risk averse. Supporting desistance by helping people with sexual convictions plan out strategies to achieve a non-offending life, while deemed necessary, is secondary to MOSOVOs. They report not having the desire, resources, or training to work in this way (Kewley, 2017; Nash, 2016) despite most

RSOs assessed by MOSOVOs as being low to medium risk (Kewley *et al.*, 2020) and of whom very few go on to reoffend (Hanson, 2018).

Kemshall's "4 Pillars of Risk Management" (*Public Protection Group. Risk of serious harm guidance 2020*, 2020) require MOSOVOs to develop risk management plans that include rehabilitative strategies to help a person successfully desist from future offending. However, helping a person develop a positive future self through, for example, seeking employment, moving house, and starting new relationships arguably requires a 'welfare' rather than a 'control-orientated' role (Blagden *et al.*, 2016). This support is a potential problem for MOSOVOs as their work is fundamentally framed around the notion that people with sexual convictions are dangerous to the public (Nash, 2019). Indeed, when carrying out home visits, the College of Policing advise MOSOVOs to "always adopt an investigative approach and be aware that offenders could potentially make convincing attempts to befriend and manipulate those who are responsible for managing them" (2020a). Such distrust or "respectful scepticism" will undoubtedly assist in the drive to ensure compliance, monitor and manage risk, detect offences, and gather intelligence. However, it is unlikely to create a safe space where the welfare of people is considered, and new identities can be fostered, developed, and tested, despite these being essential for assisting desistance (Mann *et al.*, 2019; Ricciardelli, 2018). Instead, MOSOVOs focus solely on surveillance and risk management with limited effort to help manage identity and reduce social barriers; this means relapse and persistence are likely (Ricciardelli, 2018). This focus is a particular concern, as "interventions designed solely to control and manage behaviours should be avoided, as they do not support motivation to change" (Villeneuve *et al.*, 2021, p. 92). Despite efforts, MOSOVOs are caught between correctional policy that provides a moral code to protect the public (prevent and manage risk) and professional norms and values that endorse the belief that people have the autonomy to change (Ward *et al.*, 2021).

As set out in legislation, this correctional policy is predominantly designed to punish the offender and protect the public. While there are provisions to rehabilitate the 'offender', many sentences and sanctions issued to people with sexual convictions serve as a barrier to reintegration and increase stigmatisation. Indeed, less visible prospective sanctions and sentences are issued a priori with the aim of preventing an offence (Padfield, 2017); these often run beyond prison and community sentences. One such example, sometimes referred to as the "sex offender register", is the notification requirement. Those convicted of any sexual offence listed in Schedule Three of the Sexual Offences Act 2003 are automatically required to notify authorities. Factors including the type of offence, type and length of sentence given, determine notification periods, ranging between two years to life. Notification requires people to provide information to the police, such as their date of birth, national insurance number, offence details, home address(es), passport, bank/credit card details, information on children they reside with, and intentions to travel outside of the UK. Information is annually registered; unless there are changes in circumstance, these must be made within three days of the change occurring (Beard, 2021). In addition to notification requirements, the Anti-social Behaviour, Crime and Policing Act 2014 means people may also be subjected to Sexual Harm Prevention Orders (SHPO). This prohibits people from doing a specific act, such as engaging in certain forms of employment, or particular activities on the internet. Likewise, Sexual Risk Orders (SRO) can also impose restrictions, including contact with a particular child or person over the internet or preventing them from visiting a particular place.

Similarly, the 'Child Sex Offender Disclosure Scheme' is designed for the public to ask for disclosure (for a record of child sex offences) about an individual they are concerned about who has contact with a child (Criminal Justice Act 2003). Kemshall et al. (2010), who were responsible for an early evaluation of the scheme, found that of the RSOs interviewed, the most common initial reaction was anxiety about negative reactions from their community/communities. While registries provide comfort to the general public, little evidence exists to demonstrate that these sanctions prevent further offending (Levenson *et al.*, 2016). Instead, a person with a conviction for a sexual offence could live in the community for decades without ever committing any other type of offence, yet will still be subject to criminal justice 'rituals' (Maruna, 2011), further reinforcing the identification and label of the 'sex offender.'

Formal agents such as MOSOVOs spend time in people's homes during home visits, monitoring court-ordered conditions, and developing and implementing risk management plans. Thus, they are well positioned to assist desistance by helping people manoeuvre this complex transition (Villeneuve *et al.*, 2021). However, they are inevitably compromised when attempting to effectively assist desistance because they are required to enforce such legislation. While many MOSOVOs work with people convicted of sexual offences in a respectful non-stigmatising manner, they face something of a 'civil purgatory' (Henley, 2018), as much of the work of MOSOVOs requires the formal reinforcement of a label by officially documenting the person as a "Registered Sex Offender". Consequently, the development of achievable, desirable, and meaningful new identities is likely to be compounded by punitive and pervasive restrictions and requirements, enforced by MOSOVOs. This is likely to cause unintended consequences for developing trusting relationships with MOSOVO and severely interrupt the desistance process (Willis, 2018).

Given such legislative restrictions and assumed negative attitudes held by MOSOVOs, there are consequences for a practice that requires the promotion of desistance. In particular, MOSOVOs, as offender managers, face structural and role barriers preventing them from promoting protective factors such as building individual strengths and strengthening pro-social bonds. For example, when managing risk, MOSOVOs can draw on a plethora of restrictive interventions or controls. Examples include residing at approved premises, home visits, restrictions on associations/residence/movement/activities, curfew checks, tagging, satellite tracking, covert surveillance, and use of ANPR (College of Policing, 2020b). However, the options for strategies that support change are far fewer, less specific, and not even available/suitable for all. Interventions that are deemed constructive include: attending accredited programmes (these are only available for people subject to licence or community order conditions); sharing information with appropriate agents/agencies, the provision of diversion activities such as employment and psychological or psychiatric input; and the use of support groups in the community (College of Policing, 2020b). To support the planning of appropriate selection and implementation of interventions, MOSOVOs use a structured risk management tool known as the Active Risk Management System (ARMS) (Kewley & Blandford, 2017). ARMS requires MOSOVOs to consider and evaluate risk and protective factors so that risk management plans help prevent further offending by drawing on restrictive and constructive interventions (College of Policing, 2020b).

Most criminal justice agencies support the adoption of a risk management approach that considers both risk and strengths. For example, the *HMPPS Approach to the Management and Rehabilitation of People convicted of Sexual Offences* (2021) outlines a strengths-based approach that helps agents overcome and reduce risk, and further, develop and promote a person's strengths. This approach draws on the biopsychosocial model of behaviour (Carter & Mann, 2016), which requires that offender managers help people in three domains: building biological capability (e.g., understanding neurological differences in people, teaching new skills), developing psychological capability (e.g., challenging offence-related thinking, exposure to new ways of thinking, teaching problem-solving or emotional coping skills), and strengthening peoples social capability (e.g., teaching intimacy or relationship skills, helping develop new relationships, helping find meaningful employment). However, a recent appraisal of MOSOVO ARMS assessments found overwhelmingly poor results, with inconsistencies between which risk and protective factors were identified and subsequent risk management plans, as well as a failure by assessors to provide meaningful actions to support the risks identified (Kewley *et al.*, 2020).

In a context such as this, assisted desistance is unlikely, as stigmatised interactions in which offending identities are reinforced, and a focus on managing risk and monitoring compliance are likely to stimulate a golem effect that reinforces offender labels (Mann *et al.*, 2019; Stout, 2018). The process of desistance is collaborative; while it requires individual and internal changes from within the person desisting, it also requires social and structural changes.

MOSOVOs must be supported, and also willing, to assist desistance for people with sexual convictions in two key areas: first, to help them develop an alternative (non-offending) lifestyle and identity, and second, to help them detail *how* this could be achieved (King, 2013), by providing safe, non-stigmatising and stable interactions that not only explore a future possible self but sets goals, and consider appropriate 'hooks for change' (Giordano, 2016). Successful desistance journeys are those that collaboratively respond to the unique and diverse needs of people. Unfortunately, in the current highly-politicised public protection context in which MOSOVOs operate, significant tensions exist for those attempting to assist the desistance process for people with sexual convictions.

References

Anti-social Behaviour. 2014. *Crime and Policing Act.* https://www.legislation.gov.uk/ukpga/2014/12/contents/enacted

Bailey, D. J., & Klein, J. L. (2018). Ashamed and alone: Comparing offender and family member experiences with the sex offender registry. *Criminal Justice Review, 43*(4), 440–457. https://doi.org/10.1177/0734016818756486

Beard, J. (2021). Registration and management of sex offenders. House of Commons Library (Number 5267). https://commonslibrary.parliament.uk/research-briefings/sn05267/

Blagden, N., Winder, B., & Hames, C. (2016). They treat us like human beings—Experiencing a therapeutic sex offenders prison: Impact on prisoners and staff and implications for treatment. *International Journal of Offender Therapy and Comparative Criminology, 60*(4), 371–396. https://doi.org/10.1177/0306624X14553227

Braithwaite, J. (1989). *Crime, shame and reintegration.* Cambridge University Press.

Bronfenbrenner, U. (1977). Toward an experimental ecology of human development. *American Psychologist, 32*(7), 513–531. https://doi.org/10.1037/0003-066X.32.7.513

Buchanan, M., & Krohn, M. D. (2020). Does it matter if those who matter dont mind? Effects of gang versus delinquent peer group membership on labeling processes. *Criminology, 58*(2), 280–306. doi:https://doi.org/10.1111/1745-9125.12237

Carter, A. J., & Mann, R. E. (2016). Organising principles for an integrated model of change for the treatment of sexual offending. In D. P. Boer

(Ed.), *The Wiley handbook on the theories, assessment and treatment of sexual offending* (pp. 359–381).

Christensen, L., Rayment-McHugh, S., McKillop, N., Cairns, N., & Webster, J. (2021). Understanding what works in the police management of child sex offenders in the community. *The Police Journal*, 0032258X211018791. https://doi.org/10.1177/0032258X211018791

Clarke, R. V., & Cornish, D. B. (1985). Modeling offenders decisions: A framework for research and policy. *Crime and Justice, 6*, 147–185. https://doi.org/10.1086/449106

College of Policing. (2020a). *Major investigation and public protection. Home visits.* https://www.app.college.police.uk/app-content/major-investigation-and-public-protection/managing-sexual-offenders-and-violent-offenders/home-visits/

College of Policing. (2020b*). Major investigation and public protection. Introduction to managing sexual offenders and violent offenders.* https://www.app.college.police.uk/app-content/major-investigation-and-public-protection/managing-sexual-offenders-and-violent-offenders/#offender-management

Cornish, D. B., & Clarke, R. V. (1987). Understanding crime displacement: An application of rational choice theory. *Criminology, 25*(4), 933–948. https://doi.org/10.1111/j.1745-9125.1987.tb00826.x

Cresswell, C. (2020). Why would you choose to study sex offenders?: Assisted desistance and reintegration of perpetrators of sexual harm. *Irish Probation Journal*, 17, 63–87. https://www.pbni.org.uk/wp-content/uploads/2020/11/IPJ-Vol-17-full-pdf-compressed-002...pdf#page=65

Criminal Justice Act. 2003. C.44. https://www.legislation.gov.uk/ukpga/2003/44/contents

Criminal Justice and Court Services Act 2000. C. 43. https://www.legislation.gov.uk/ukpga/2000/43/2000-11-30.

Farmer, M., McAlinden, A. M., & Maruna, S. (2015). Understanding desistance from sexual offending A thematic review of research findings. *Probation Journal, 62*(4), 320–335. https://doi.org/10.1177/0264550515600545

Farrall, S., Bottoms, A., & Shapland, J. (2010). Social structures and desistance from crime. *European Journal of Criminology, 7*(6), 546–570. https://doi.org/10.1177/1477370810376574

Farrall, S., Hunter, B., Sharpe, G., & Calverley, A. (2014). *Criminal careers in transition: The social context of desistance from crime*. Oxford University Press.

Flynn, N. (2017). *Criminal behaviour in context: Space, place and desistance from crime*. Taylor & Francis.

Giordano, P. C. (2016). Mechanisms underlying the desistance process: reflections on 'A theory of cognitive transformation'. In J. Shapland, D. Farrington, & A. Bottoms (Eds.), *Global Perspectives on Desistance* (pp. 27–43). Routledge.

Giordano, P. C., Cernkovich, S. A., & Rudolph, J. L. (2002). Gender, crime, and desistance: Toward a theory of cognitive transformation. *American Journal of Sociology, 107*(4), 990–1064. https://doi.org/10.1086/343191

Giordano, P. C., Longmore, M. A., Schroeder, R. D., & Seffrin, P. M. (2008). A life-course perspective on spirituality and desistance from crime. *Criminology, 46*(1), 99–132. https://doi.org/10.1111/j.1745-9125.2008.00104.x

Glueck, S., & Glueck, E. (1950). *Unraveling juvenile delinquency*. Harvard University Press.

Grady, M. D., Levenson, J. S., Mesias, G., Kavanagh, S., & Charles, J. (2019). I cant talk about that: Stigma and fear as barriers to preventive services for minor-attracted persons. *Stigma and Health, 4*(4), 400. doi:https://doi.org/10.1037/sah0000154

Hanson, R. K. (2018). Long-Term Recidivism Studies Show That Desistance Is the Norm. *Criminal Justice and Behavior, 45*(9), 1340–1346. https://doi.org/10.1177/0093854818793382

Hanson, R. K., Harris, A. J., Letourneau, E., Helmus, L. M., & Thornton, D. (2018). Reductions in risk based on time offence-free in the community: Once a sexual offender, not always a sexual offender. *Psychology, Public Policy, and Law, 24*(1), 48. https://doi.org/10.1037/law0000135

Healy, D. (2013). Changing fate? Agency and the desistance process. *Theoretical Criminology, 17*(4), 557–574. doi:https://doi.org/10.1177/1362480613494991

Henley, A. (2018). Mind the gap: Sentencing, rehabilitation and civic purgatory. *Probation Journal, 65*(3), 285–301. https://doi.org/10.1177/0264550518776773

Hirschi, T. (1969). *Causes of delinquency*. University of California Press.

Hirschi, T., & Gottfredson, M. (1983). Age and the explanation of crime. *American Journal of Sociology, 89*(3), 552–584.

HM Prison & Probation Service. (2021). The HMPSS Approach to the Management and Rehabilitation of People Convicted of Sexual Offences. *Version 1.2.* https://www.gov.uk/government/publications/the-hmpps-approach-to-the-management-and-rehabilitation-of-people-convicted-of-sexual-offences

Kemshall, H., & Hilder, S. (2020). Effective Systems and Processes for Managing Violent Offenders in the United Kingdom and the European

Union. In J. S. Wormith, L. A. Craig, & T. E. Hogue (Eds.), *The Wiley Handbook of What Works in Violence Risk Management* (pp. 485–503).

Kemshall, H., Wood, J., Westwood, S., Stout, B., Wilkinson, B., Kelly, G., & Mackenzie, G. (2010). Child Sex Offender Review (CSOR) Public Disclosure Pilots: a process evaluation -2nd edition. Research Report 32. https://assets.publishing.service.gov.uk/government/uploads/system/uploads/attachment_data/file/214959/horr32c.pdf

Kewley, S. (2017). Policing people with sexual convictions using strengths-based approaches. *Journal of Criminal Psychology, 7*(3), 168–182. https://doi.org/10.1108/JCP-09-2016-0026

Kewley, S., & Blandford, M. (2017). The development of the active risk management system. *Journal of Criminal Psychology, 7*(3), 155–167. https://doi.org/10.1108/JCP-10-2016-0034

Kewley, S., & Brereton, S. (forthcoming). Public Protection. In L. Burke, N. Carr, E. Cluley, S. Collett, & F. McNeill (Eds.), *Reimagining Probation Practice*.

Kewley, S., Osman, S., & McGuinness, Á. (2020). How well do police specialists risk assess Registered Sexual Offenders? *Journal of Sexual Aggression, 26*(3), 302–315. doi:https://doi.org/10.1080/13552600.2019.1628315

King, S. (2013). Assisted desistance and experiences of probation supervision. *Probation Journal, 60*(2), 136–151. https://doi.org/10.1177/0264550513478320

King, S. (2014). *Desistance Transitions and the Impact of Probation*. Routledge.

Kras, K. R. (2019). Can Social Support Overcome the Individual and Structural Challenges of Being a Sex Offender? Assessing the Social Support-Recidivism Link. *International Journal of Offender Therapy and Comparative Criminology, 63*(1), 32–54. https://doi.org/10.1177/0306624x18784191

Laub, J. H., & Sampson, R. J. (2001). Understanding Desistance from Crime. *Crime and Justice, 28*, 1–69. http://www.jstor.org/stable/1147672

LeBel, T. P., Burnett, R., Maruna, S., & Bushway, S. (2008). The chicken and egg of subjective and social factors in desistance from crime. *European Journal of Criminology, 5*(2), 131–159. https://doi.org/10.1177/1477370807087640

Levenson, J. S., Grady, M. D., & Leibowitz, G. (2016). Grand Challenges: Social justice and the Need for Evidence-based Sex Offender Registry Reform. *Journal of Sociology. & Social Welfare, 43*(2), 3–38. https://heinonline.org/HOL/LandingPage?handle=hein.journals/jrlsasw43&div=20&id=&page=

Mann, N., & Lundrigan, S. (2021). Dynamic assessment of registered sexual offenders: the national practitioner perspective on the use of the Active risk management System. *Policing and Society, 31*(10), 1199–1216. https://doi.org/10.1080/10439463.2021.1873324

Mann, N., Devendran, P. N., & Lundrigan, S. (2019). You're never really free: Understanding the barriers to desistance for registered sexual offenders in the community. *Criminology & Criminal Justice, 21*(2), 206–223. https://doi.org/10.1177/1748895819853861

Mann, N., Devendran, P., & Lundrigan, S. (2018). Policing in a Time of Austerity: Understanding the Public Protection Paradox through Qualitative Interviews with Police Monitoring Officers. Policing. *A Journal of Policy and Practice, 14*(3), 630–642. https://doi.org/10.1093/police/pay047

MAPPA Guidance. (2021). National Offender Management Service Offender Management and Public Protection Group. https://mappa.justice.gov.uk/MAPPA/view?objectID=108075909

Maruna, S. (2001). *Making good: How ex-offenders reform and reclaim their lives*: American Psychological Association.

Maruna, S. (2011). Reentry as a rite of passage. *Punishment & Society, 13*(1), 3–28. https://doi.org/10.1177/1462474510385641

Maruna, S., & Mann, R. (2019). Reconciling Desistance and What Works. *HM Inspectorate of Probation Academic Insights 2019/1.* Retrieved from https://www.justiceinspectorates.gov.uk/hmiprobation/wp-content/uploads/sites/5/2019/02/Academic-Insights-Maruna-and-Mann-Feb-19-final.pdf

Matza, D. (1964). Position and behavior patterns of youth. In *Handbook of modern sociology* (pp. 191–215). Rand McNally.

McAlinden, A.-M., Farmer, M., & Maruna, S. (2016). Desistance from sexual offending: Do the mainstream theories apply? *Criminology & Criminal Justice, 17*(3), 266–283. doi:https://doi.org/10.1177/1748895816670201

McCartan, K., & Gotch, K. (2020). International Approaches to the Management of Perpetrators of Sexual Harm Policy: Preventative, Practical, or Political? *The Wiley Handbook of What Works with Sexual Offenders: Contemporary Perspectives in Theory, Assessment, Treatment, and Prevention*, 441–454. John Wiley & Sons Ltd.

McCartan, K., Kemshall, H., & Tabachnick, J. (2015). The construction of community understandings of sexual violence: Rethinking public, practitioner and policy discourses. *Journal of Sexual Aggression, 21*(1), 100–116. https://doi.org/10.1080/13552600.2014.945976

McNeill, F. (2016). The collateral consequences of risk. In C. Trotter, G. McIvor, & F. McNeill (Eds.), *Beyond the risk paradigm in criminal justice* (pp. 143–157).

McNeill, F. (2019). *Pervasive punishment: Making sense of mass supervision.* Emerald Group Publishing.

Mullins, E., & Kirkwood, S. (2019). Co-authoring desistance narratives: Analysing interactions in groupwork for addressing sexual offending. *Criminology & Criminal Justice, 21*(3), 316–333. https://doi.org/10.1177/1748895819863101

Multi-Agency Public Protection Arrangements–Annual Report. (2021). *2020/21. Ministry of Justice Statistics Bulletin.* https://mappa.justice.gov.uk/MAPPA/view?objectID=31271312

Nash, M. (2019). MAPPA: Sex offenders and managing the other in the community. In A. Pycroft & D. Gough (Eds.), *Multi-Agency Working in Criminal Justice: Theory, Policy and Practice* (2 ed., pp. 7588). Policy Press.

Nash, M. R. (2016). Scum Cuddlers: police offender managers and the sex offenders register in England and Wales. *Policing and Society, 26*(4), 411–427. https://doi.org/10.1080/10439463.2014.942855

Ouellet, F. (2019). Stop and go: Explaining the timing of intermittency in criminal careers. *Crime & Delinquency, 65*(5), 630–656. https://doi.org/10.1177/0011128717753114

Padfield, N. (2017). Preventive sentencing. In K. McCartan & K. H. (Eds.), *Contemporary Sex Offender Risk Management, Volume I: Perceptions. Palgrave Studies in Risk, Crime and Society* (pp. 89–114). Palgrave. https://doi.org/10.1007/978-3-319-63567-5_6

Paternoster, R., & Bushway, S. (2009). Desistance and the feared self: Toward an identity theory of criminal desistance. *The Journal of Criminal Law and Criminology, 99*(4), 1103–1156. https://doi.org/0091-4169/09/9904-1103

Piquero, A. R., Monahan, K. C., Glasheen, C., Schubert, C. A., & Mulvey, E. P. (2013). Does Time Matter? Comparing Trajectory Concordance and Covariate Association Using Time-Based and Age-Based Assessments. *Crime & Delinquency, 59*(5), 738–763. doi:https://doi.org/10.1177/0011128712459491

Public Protection Group. (2020). *Risk of Serious Harm Guidance 2020.* https://assets.publishing.service.gov.uk/government/uploads/system/uploads/attachment_data/file/897166/rosh-guidance-2020.pdf

Ricciardelli, R. (2018). Parolee Perceptions of Case Management Practices during Reintegration. *Victims & Offenders, 13*(6), 777–797. https://doi.org/10.1080/15564886.2018.1476997

Sampson, R. J., & Laub, J. H. (2003). *Shared beginnings, divergent lives: Delinquent boys to age 70*. Harvard University Press.

Sexual Offences Act. 2003. C.42. https://www.legislation.gov.uk/ukpga/2003/42/contents

Stout, B. (2018). Desistance–lessons learned for managing violent and sexual offenders. In J. L. Ireland, C. Ireland, & P. Birch (Eds.), *Violent and Sexual Offenders* (pp. 520–533). Routledge.

Thomas, T. (2016). *Policing sexual offences and sex offenders*. Springer.

Thomas, T., & Marshall, D. J. (2021). *The Sex Offender Register: Politics, Policy and Public Opinion*. Routledge.

Villeneuve, M.-P., Dufour, F., I, & Farrall, S. (2021). Assisted desistance in formal settings: A scoping review. *The Howard Journal of Crime and Justice, 60*(1), 75–100. https://doi.org/10.1111/hojo.12396

Ward, T., Durrant, R., & Dixon, L. (2021). The classification of crime: Towards pluralism. *Aggression and Violent Behavior, 59*, 101474. https://doi.org/10.1016/j.avb.2020.101474

Weaver, B. (2016). *Offending and desistance: The importance of social relations*. Routledge.

Weaver, B. (2019). Understanding desistance: A critical review of theories of desistance. *Psychology, Crime & Law, 25*(6), 641–658. https://doi.org/10.1080/1068316X.2018.1560444

Willis, G. M. (2018). Why call someone by what we don't want them to be? The ethics of labeling in forensic/correctional psychology. *Psychology, Crime & Law, 24*(7), 727–743. https://doi.org/10.1080/1068316X.2017.1421640

Winder, B., Bladgen, N., Lievesley, R., Dwerryhouse, M., Kitson-Boyce, R., & Elliot, H. (2020). *UK National Evaluation of Big Lottery Funded Circles of Support and Accountability*. https://circles-uk.org.uk/wp-content/uploads/2021/10/SOCAMRU_BL_final_report_1_April_2020_v82.pdf

3

Negotiating Identity as a 'Sex Offender' on Release from Prison and the Negative Impact of Restrictions

Elizabeth Hill and Nicholas Blagden

Introduction

Currently, over 15,000 individuals are imprisoned for sexual offences in England and Wales (Ministry of Justice, 2025). Sentences range from community orders to life imprisonment (Sentencing Council, 2014), and annually, approximately 2800 individuals with sexual convictions are released into the community (FOI response, 2021). The overwhelming majority (>99%) of individuals in prison for sexual offences are men, and most will be released into the community.

Given the increase in the number of people in prison for a sexual offence conviction, it is notable that holistic and well-rounded support for people when they are released from prison has remained limited.

E. Hill (✉)
His Majesty's Prison & Probation Service, London, UK
e-mail: elizabeth.b.hill@justice.gov.uk

N. Blagden
University of Derby, Derby, UK
e-mail: n.blagden@derby.ac.uk

N. Blagden et al. (eds.), *Sexual Crime and Community Reintegration*, Sexual Crime,
https://doi.org/10.1007/978-3-032-06034-1_3

The criminal justice and societal response remains dominated by perceptions of risk underpinning the public (or community) protection model (Kemshall & Maguire, 2001). This model prioritises control over change and is characterised by the utilisation of (i) risk assessment tools; (ii) civil injunctions and restrictive orders; (iii) post-custody licence and registration conditions; (iv) multi-agency public protection arrangements; and (v) offending behaviour programmes. Kewley and Brereton (2022) describe policies and practices aligned with the public protection approach as being characteristic of 'penal excess'. Unsurprisingly and consequentially, community experiences are shaped by both stigma and shame (Cubellis et al., 2019). As such, re-entering the community is challenging for people with sexual convictions due to personal, social, and societal obstacles (Bows & Westmarland, 2018; Mears & Cochran, 2015, Rydberg, 2018).

Desistance literature suggests that punitive restrictions create obstacles to securing or developing protective factors and hinder the transition from prison to the community (Brown et al., 2007; Tolson & Klein, 2015). Indeed, a key factor for 'making good' is having a place in society. Yet, as Bows and Westmarland (2018) argue, a sexual conviction can lead to isolation, limited contact with family and friends, and mental distress, which are risk factors for reoffending (Grossi, 2017). Tovey et al. (2022) found that employment was central to people's identities and social status, as work is construed as a way of contributing and belonging to something. Research by Tovey et al. (2022) found that participants experienced an acute sense of loss at being denied opportunities to work and that there was a longing for pro-social relationships with peers. Social interactions with such peers can be a protective factor against recidivism and are of benefit to a person's general psychological well-being and social capital (Harris et al., 2020). While there are challenges for employers, including the stigma of a sexual offence, fears surrounding risk, and what other workers may think/do, there remains a general reluctance to hire people with sexual convictions (Tovey et al., 2022).

Current Study

The current study aims to build upon previous literature by further exploring the experiences of men with sexual convictions attempting to reintegrate into the community, thus adding to the knowledge base about the UK experience specifically. The interviews placed particular emphasis on how being on a 'sex offender' register impacted self-identity and self-perception. Participants were also asked for their perspectives on reintegration, desistance, the impact of registration, and managing their own identity.

Method

Participants

Nine participants took part in this study (Table 3.1 for details). All were men who had been convicted of a sexual offence (MCOSOs), served a prison sentence for this, and had since been released back into the community. Access to participants was granted by the London National Probation Services Research Committee and the Safer Living Foundation, a charity which supports those convicted of or at risk of committing sexual offences. Participants were recruited via Probation Officers and other relevant mental health or support professionals working with this population. Participants were aged between 24 and 62, with a mean age of 42. Their offences included possession of indecent images of children, attempted rape, and rape. Their victims included children and adults, both known and unknown.

Data analytical procedure

This study used interpretive phenomenological analysis (IPA) to qualitatively analyse participants' accounts. IPA is an idiographic approach concerned with exploring individuals' lived experiences and the meanings they attribute to those experiences (Smith & Eatough, 2007). The

Table 3.1 Participant information

Participant	Age	Conviction	Time in Prison (years)	Time on release
1	52	2 counts of rape, against stranger adult females	18 (IPP)	Unknown
2	26	Attempted rape against adult known woman	2.5	Unknown
3	24	Sexual assault and attempted rape–stranger-adult woman	4.5	4 months
4	32	One count rape and nine counts inciting a minor to engage in sexual activity–male victim	7.2	1 year 4 months
5	62	Indecent assault against male child and female child	8	4 years
6	58	Sexual assault, committing a sexual act in the presence of a child, inciting a child to engage in sexual activity	9	1 year
7	35	Breach SOPO	1.5	1 year
8	35	Rape, 1x sexual assault, 3 x digital penetration, known female child (cousin), 12 years old	4	1 year
9	55	Indecent images. Previous convictions for attempted rape and indecent assault against known child victims	9	2.5 years

analytical procedure entails a double hermeneutic; with the researcher "trying to make sense of the participants trying to make sense of their world" (Smith & Osborn, 2003, p. 51). Interview transcripts were read multiple times to allow the researcher to immerse themselves in the data (Smith & Osborn, 2003), and descriptive, conceptual, and linguistic coding was utilised in a process where recurring patterns of meaning within the data were developed into subordinate themes and clustered under broader superordinate theme headings. The final themes were representative of the sample. A form of inter-coder agreement was used as a verification procedure to check coding of qualitative data (see De Wet & Erasmus, 2005; Smith, 2015). In qualitative research this occurs when two or more researchers code the exact same data independently and check for consistency across coders (De Wet & Erasmus, 2005). This chapter unpacks one main superordinate theme 'Negotiating Identity as a "sex offender" and associated subthemes.

Negotiating Identity as a 'Sex Offender'

One of the most salient superordinate themes for participants was negotiating their identity: who they were, who they are now and how others see them. Maruna (2001) stated that those who are desisting may reject their offending identity and consider themselves to have been 'good' all along ('making good'); they can then reconcile their past behaviour with who they are now, and still maintain a positive identity. Some participants were still facing significant difficulty with working out where their offending fit with their current identity, suggesting they needed more support with this.

Battling with the Label of 'Sex Offender'

Participants explained how they saw themselves, how others saw them, and considered whether the label 'sex offender' fit with this.

> "I view myself as somebody that's committed a sex offence…which happened to be against (pause) children…Ok, so that's that…right? I don't view myself as a sex offender and I don't view myself as a paedophile…" (Ptpt 6, extract 1)

> "Well yeah, you know I am (a 'sex offender')…yeah you know I committed a crime, no point in hiding it or sugar coating it…if you

> want to get the treatment, you know, you want to try and help yourself...you've got to come to terms with what you've done and...what you are...if you don't come to terms with what you've done and what you are, no matter how much you do, it just won't help...you're still in denial...
>
> Interviewer: Yeah that's a good point, but would you think of yourself as someone who's committed a sexual offence or someone who is a 'sex offender' or was a 'sex offender'?
>
> Ptpt: Er, I would say someone that has committed a sexual offence and a recovering sex offender...you know it's like alcoholism...you're never fully recovered, you're still recovering" (Ptpt 7, extract 2)

Participants battled with this issue, some going back and forth on whether they considered themselves to be 'sex offenders', still, if they acknowledged that label, or what it meant for them. Participants 6 and 7 distinguished between someone who had committed a sexual offence and someone who was a 'sex offender'. Distancing themselves from the label (Winder & Gough, 2010) may have helped avoid internalising the shame associated with their offences. Participant 7, however, then referred to himself as a 'recovering sex offender', indicating that 'sex offender' was still part of his identity, albeit one 'in recovery'.

> "Well I'm a fucking rapist, aren't I?
>
> Interviewer: Is that how you think of yourself?
>
> Ptpt: Erm sometimes yeah...sometimes yeah, if I'm really negative towards myself do you know what I mean, yeah, and I think it separates me, I mean I am different do you know what I mean, I do have a lot more conditions on my life do you know what I mean...you know I have to do this thing, you know you read the papers do you know what I mean and you get your view on what sex offenders are and stuff like this, you speak to people, you know someone told me that if [their] son grew up to be a sex offender they'd kill 'em, hearing stuff like this" (Ptpt 8, extract 3).

Participant 8 had particularly struggled with self-identity and what that meant to him, and expressed a lot of negative emotions about himself and his situation. He felt separated from others by the label, including the media perception of 'sex offenders' and the restrictions placed on him. This separation made it difficult not to think about himself negatively, as he was highly aware of others' negative perceptions.

> "If I continually think of myself as a rapist I'm going to feel bad about it but at the end of the day I did commit rape so I am a rapist…so I don't really mind…but yeah someone who's committed a sexual offence is probably the best way of saying it do you know what I mean…cos I wouldn't consider myself a paedophile or a rapist moving forward…or anything like that, you know I would consider myself just somebody who's done something that was fucking..unbelievably…er anyway do you know what I mean…so I don't, but I understand that those views can and do do have an impact…you know if I really, if I sit and tear myself apart do you know what I mean, calling myself a rapist and you know all things that come with it…that is going to impact on me, so I don't do that…but yeah someone who's committed a sex offence is probably the best way of labelling myself" (Ptpt 8, extract 4)

Participant 8 was attempting to be pragmatic about the labels used for himself and not let them affect him. However, he clearly found this difficult, particularly when, as seen in extract 4, he could not bring himself to describe his offence out loud and had to move on. Participant 6 explained how being on the register made him different from other people and admitted that he would not accept himself if in someone else's shoes.

> "I find it difficult to understand that there are people about that erm don't ignore what I've done…Ok? But don't actually see that as the whole picture of me…they take me for me…because I wouldn't and I I I'll just talk straight here, if it was me, if I was in their position…I would have nothing to do with me…yeah yeah so I would have nothing to do with me so it's a good job there are (laughs) people about that are actually erm I don't know more forgiving more understanding…erm because otherwise (laughs) I'd have no-one to talk to, would I" (Ptpt 6, extract 5)

As seen in extract 5, he did not consider himself worthy of acceptance, but expressed gladness that others did, and recognised his view had disadvantages. Those he referred to as being understanding were professionals. He was perhaps choosing not to focus on how their jobs required interacting with him positively, but nonetheless, their acceptance was important to him.

> "Well it makes you different because you have to sign and they don't…when you move, so yes from that perspective, but in general thinking, no…no, it's only because of the things that you have to do…yeah that makes it different, but I don't feel any different because I'm on the register, from yourself for example" (Ptpt 9, extract 6)

In contrast, Participant 9 emphasised how there were no psychological differences between him and others, only the practical aspects of being on the register.

> "Er I didn't realise straight away about myself but I recognise now, and before then, that I'm always going to have sexual urges towards children…but I'm learning how to cope with them, deal with them…you know, cos you'll always have those urges…against children, I'll always have those urges, so you've gotta recognise them" (Ptpt 7, extract 7)

> "You know because I had that interest (in children) …and it doesn't go away…it's being able to control it" (Ptpt 9, extract 8)

Participants 7 and 9 referred to their risk of offending as ongoing, despite desisting and not planning to reoffend. Participant 7 referred to always having urges 'against' children; this could indicate an understanding of the link between his urges and potential sexual violence, or that he saw himself and children on opposite sides of a battle he was still trying to win. Stigma and restrictions which affect integration can communicate a message of 'irredeemability' and remaining at risk of reoffending always (Kemshall, 2008). Seeing risk as life-long or ongoing could be contrary to building a new desisting identity, as it assumes ongoing 'deviant' interests. However, it could instead be seen as complementary: sexual interests are part of the individual's history and makeup, however, they are managing these effectively and therefore desisting.

> "You live a life of misery, not misery, loneliness…I'm glad I'm not 20 or 30…Do you know what I mean? Cos they've got no chance, they can never have a kid…Cos they think they're gonna fiddle with their own kid, it's just…But you go and murder someone, you can get married and have kids yourself…But you know, they took a life, it's but it happens doesn't it, they come out after, we let 'em out and murderers reoffend you know what I mean…so there's good people who wants to try and go straight and there's bad people who don't wanna go straight…That's why I think they should look at different people and say…Right we're gonna give him some slack because he's doing quite well" (Ptpt 5, extract 9)

Participant 5 distinguished between 'good people' who want to desist, and 'bad people' who continue offending, and in doing so established himself as one of the 'good' ones who wanted to change. Hochstetler et al. (2010) similarly found that those who had committed violent offences tended to define others as 'authentically' violent, but saw themselves as having acted violently for other reasons.

> "They call us sex offenders but you don't have sex no more because of it, cos you can't get a relationship, unless you go and get a prostitute or something…which you know, that's even breaking your licence condition…so you're living like a monk sort of thing…cos of the fear …I think to be honest with you, I've been inside and out and a lot of sex offenders who reoffended couldn't get relationships so they want sex so they go back to touching a kid up
>
> Interviewer: Back to offending?
>
> Participant: Back to offending, because we're all like animals, we all wanna have a sex life…" (Ptpt 5, extract 10)

Participant 5 focused on the impact of his conviction and the label of 'sex offender' on his ability to have future sexual relationships. He feared breaking his licence conditions by seeking a sexual relationship but also believed that having no opportunities for sex could lead to reoffending. This suggests a worry that whatever he decided, there was a risk of reoffending, a feared possible self (Markus & Nurius, 1986). This way of thinking could potentially absolve those who reoffended of responsibility for their actions; if they are animals, could they truly control themselves? More than that, he felt a loss of identity due to not being able to have sex anymore, which he saw as a basic human need he was no longer entitled to. It is notable that he referred to everyone as 'animals' here; while the context is different, the same participant appreciated being treated as a 'human being' (extract not included here). Perhaps dividing those who had not changed (animals) and those who had, like himself (human beings), was another way to separate himself from his past actions, though he worried about reverting to his 'animal' instincts.

> "(Regarding people finding out) Erm I wouldn't say concerns, I just think it's irrelevant erm but you know I've got a few mates from work that we all go out for a drink or a meal as a group…and I don't feel that I cou-, that they need to know…erm, you know that's something that I've done in my past…they've only got to know me from who I am now…erm so I've chose not to bring my past up…
>
> Interviewer: do you ever worry that they would find out even if you don't disclose?
>
> Ptpt: Oh yeah I mean it's always on my mind you know for those I haven't told, erm but I just hope that kind of by that point they kind of know me for a while…and got to know me as a person, and if it comes up I just hope they'll give me a benefit of the doubt…erm let me explain and I'll just simply say it you know that's my past…you kind of met me now… didn't feel like my past of any relevance." (Ptpt 4, extract 11)

Participant 4 had decided not to tell his new friends about his past; Winnick & Bodkin (2009) described this as a secrecy strategy, 'passing' as non-stigmatized. He appeared to begin saying he could not tell his friends, then revised it to them not needing to know. To protect against potentially losing these friends, it makes sense to tell himself they do not need to know, as he is a different person now. However, he did acknowledge that whether they might find out was 'always on (his) mind'. These friendships were therefore difficult to navigate, getting closer to people but across a barrier that cannot be crossed for fear of losing the relationship.

> "When I'm at work or going to work, I don't get people stopping or asking me questions…so I feel like I'm fitting in…but it does in a way like, knowing that I'm on the register…like, it's just it's just like a little tag innit…yeah I know inside but other people ain't aware of" (Ptpt 3, extract 12)

Work, for Participant 3, helped maintain a positive self-identity, even if underneath, he still felt different. Evidence for the role of work in desistance from sexual reoffending is mixed. Some MCOSOs were in employment when they offended, so work was not a protective factor for them (Harris, 2014), and in some cases will have facilitated access to victims. However, as participant 3 highlighted above, work was important in self-identity and feeling accepted. These factors may have a knock-on effect on desistance in terms of having an identity inconsistent with reoffending (Maruna, 2001).

Sex Offenders Are the Worse Cases on Earth

While Participant 5's views on sex offenders had developed over time, he still retained the belief that they, and by extension he, were the 'worse cases on earth', which caused him to struggle.

> "Unfortunately sex offenders are the worse cases on earth, people look at 'em –
>
> Interviewer: Do you think that though? Or is that just what people say?
>
> Ptpt: Everyone does, everyone, you talk to. We're the worst lot, we're worse off than anyone.

Interviewer: But I mean do you think that they're the worst case on earth?

Ptpt: I don't so much now, but I used to." (Ptpt 5, extract 13)

Participant 5's admission that he does not think of himself as the worst case 'so much now' indicates that he still does sometimes; this was another ongoing battle, to maintain a positive view of himself and not let others' perceptions influence this. There appears to be hope, however, that he was currently winning this battle more often than not. Lester (2008) stated that 'sex offender' is the 'most damning label in modern society' (p.107); the public has more negative attitudes towards those who have committed sexual offences than other offences (Willis et al., 2010).

A number of participants were clearly aware of this, and some referenced how people who have killed someone are seen less negatively than those who have committed sexual offences. Indeed, there is no register relating to violent offences or offences against intimate partners, where rates of reconviction are higher. The fact that only those with sexual offences are monitored in this way sends a message to the public, but also 'ex-offenders' themselves, that they are the 'worst' 'offenders', and need to be closely monitored so they do not reoffend. Seidler (2010) also found that Australian MCOSOs highlighted how they were the only types of 'ex-offenders' subject to a register, considering this an additional punishment. By comparing themselves to those who had killed someone, participants may also be trying to establish another outgroup, which they considered to be 'worse' than them (Hochstetler et al., 2010).

"People hate us more, I think people can tolerate a lifer who's killed someone…than someone who's touched a kid up or something, raped a woman…or someone who you know does any sex offence…people wanna burn us by the slate (sic) like witches and kill us all" (Ptpt 5, extract 14)

"Someone told me that if [their] son grew up to be a sex offender they'd kill 'em, hearing stuff like this…you know so it's like even my own mother said she'd rather I murdered (the victim)" (Ptpt 8, extract 15)

Participants 5 and 8 had internalised the idea that people would want them dead due to their offending, and that they could have done nothing worse. The comparison with witches is also interesting. It perhaps conveys not only the way in which suspected witches were treated but also the modern-day meaning of a 'witch hunt', implying a targeted attack which can be disproportionate with the actions of the person being 'hunted'.

> "So when someone says oh you're a sex offender, you're a child fiddler, you're a thing…that kind of knocks you for six…because you know you are and that'll be with you for the rest of your life…but you don't wanna be thinking (inaudible) of it especially when you're trying to get on with your life…you don't wanna keep coming back to that, I mean you can never get away from it, you can never, you can never stop saying sorry…there's a time comes when you gotta stop saying sorry...and live a better life, so you're not gonna say sorry to anyone else…yeah if people kept saying to me you're a sex offender…it would get on my tits… I don't call myself sex offender now but I am a sex offender" (Ptpt 5, extract 16)

Participant 5 illustrated the difficulties in managing his identity and his changing view of it. He was distressed at being labelled a 'sex offender' and not being able to escape this label, but also irritated that others kept relating to him in this way, even when he agreed with them. This constitutes a significant number of conflicting views to hold at one time. To Participant 5, knowing how MCOSOs are viewed, and even thinking about himself in that way to a certain extent, just made the label more difficult to deal with. Crocker and Major (1989) suggest that people can incorporate attitudes of how others see them into their view of themselves.

Fear of Identity Being Exposed

Participants were concerned about others finding out about their offending, both the general public and people closer to them; this would be a community barrier as described by Burchfield and Mingus (2008). Consistent with other research (Lasher & McGrath, 2012; Tewksbury &

Lees, 2006), participants were fearful of repercussions to their offending being discovered, even when the majority of them had not personally experienced victimisation. Information on the likelihood of reprisal is something that is important to share with MCOSOs prior to release from prison, to put their worries into context and perhaps allay some of their fears.

> "At the moment I feel like I'm exposed, I'm not far off the streets, you know the pavement, in the building...and I'm conscious, you know I've got to keep looking over my shoulder, is someone standing there...looking this direction, I have to be careful how I slide the forms back under...yeah, you know, just like you really feel dirty...nasty, you know you don't feel like a human, er you know you feel dehumanised...I've got to sort of like stake the area out, make sure no-one's around..." (Ptpt 4, extract 17)

Participant 4 focused on a specific moment of fear: having to attend the police station to sign the 'sex offender' register. While this was something he only had to do occasionally (usually once a year for those on the register), he was clear that the fear of people finding out was something that bothered him 'all the time', illustrating the disproportionate amount of time he spent worrying about being 'found out'. Feeling confident in his new identity would presumably be difficult with this fear hanging over him. It is also interesting that Participant 4 described not feeling like a human during these times; perhaps his use of language, i.e. staking the area out, and 'keeping tracks' suggests he felt more like an animal, who was being hunted.

> "You always gonna have fear that someone's gonna spot you, so that's always there...because you think oh something's gonna happen because I'm too happy, so that's always there but that's just me" (Ptpt 5, extract 18)

Participant 5 not only had a constant fear that someone would 'spot' him, but he also linked this to his happiness; if he was too happy, he would be punished by being exposed for what he had done. Evans and Cubellis (2015) suggested that registered 'sex offenders' who thought about their registration status often, even daily, also had a generally negative outlook on life and themselves.

Acceptance of Self

One specific strategy participants used, when considering how they were seen now, was accepting their past and using this as a way to keep moving forward.

> "I'm sorry for what I did, let it go (laughs), let's move on" (Ptpt 1, extract 19)

> "I don't forget where I've come from and all the stuff, er, before I go to sleep, when I wake up, I don't forget, I'm grateful for the fact that I'm out here, that I'm living again, people have given me an opportunity, another chance in life and things like that" (Ptpt 1, extract 20)

Participant 1 was asking to be allowed to move on as he regretted his offences; he had committed two rapes against strangers, so this is a difficult position to make in terms of the severity and rarity of these types of crimes, but he appreciated the positive aspects of his life and those people who had let him move on.

> "The fact is that people are what they are, they're gonna judge, they're gonna box you in, they're gonna stereotype you, they're gonna label you, they're gonna do whatever they've gotta do, that's their issue." (Ptpt 1, extract 21)

He also, however, accepted the fact that others are unlikely to see him in the same way and may continue to stereotype him. This is similar to some participants in Harris' (2014) study who had reported desisting due to their age; they did not see themselves as 'sex offenders' but had resigned themselves to continued stigma from others. They however, tended to be pessimistic about their future, whereas Participant 1 focused more on gratitude for finally being out of prison.

> "I do have bad days but I always try to look at things positively, erm, and erm I can look at it oh my god they're picking on me and this that and the other...or I can say well hold on a minute, if I've moved into an area and they've got concerns, they may not have concerns about me, but it's the welfare of people around me...Ok, and I know what I've done and I'm ashamed of what I've done, erm, but I'm still a family man...and erm children and grandchildren, and if I was in that position, then I would want to know...there's certain things in life that are not forgivable...and what I did was one of those things...right? So I got through it two ways – the first way was by just sort of, I'm upset that they don't wanna know me but I'm proud at the same time that like what they're doing is they're following...the moral code and everything else that erm that I lived." (Ptpt 6, extract 22)

Participant 6 accepted responsibility for his actions and why society would feel negatively towards him, and therefore also the need to comply with rules placed upon him. He used perspective taking to consider what he would think about someone with a sexual offence being around his family. One of his victims was a family member, but family was still important to him, and he could even see his rejection by them in a positive light. This seemed to be an effective way to deal with negative feelings about being isolated from his family, as even that was proof he had raised his family in the right way. It perhaps also helped him remember that he had lived by a 'moral code' at least at one time, and had not always been the person he was when he committed his offences.

Negative Impacts of the 'Sex Offender' Register and Restrictions

Another particularly salient theme concerned the restrictions in place following release from prison, and their significant impact on daily life.

It's a Hurdle

Participant 1 described the restrictions associated with being on the 'sex offender' register as a 'hurdle'.

> "It impacts your life obviously…it's erm one of those things that, when you're looking for jobs or anything else in life…whatever kind of thing you're trying to do, whether it's erm financially, if you're trying to look for something, if you're trying to get some insurance, if you're looking for erm erm a passport, if you're looking for anything…then you've got to declare it, and it's, it's a hurdle…it's a big hurdle." (Ptpt 1, extract 23)

A hurdle indicates a possibility of getting over these restrictions as opposed to a barrier which might prevent moving forward. However, it also implies that action is needed; overcoming the hurdles was proving difficult for Participant 1 as they were largely out of his control. This applied to 'anything', so it was a constant problem when trying to achieve things in his life.

Some participants emphasised that it was not the 'sex offender' register which had the largest impact, but the Sexual Offences Prevention Order (SOPO, now Sexual Harm Prevention Order) and licence conditions. As indicated by Participant 9, it was these restrictions that reduced his opportunities to have a normal life:

> "I got another place, a council flat, which again you know Probation and the police have to approve, they both did…and then the day before I was due to move in, the council said no, cos…the supervisor of the area decided that it was too close to a school, even though the police were happy…so that was a bit of a knock back, so it's things like that which you know, we face…and it is sometimes hard to understand when, you know, for example, the people in authorities, who you know know you claiming something and then someone who doesn't know you just do a general
>
> Interviewer: override it higher up or something?
>
> Ptpt: Yeah that's quite frustrating." (Ptpt 9, extract 24)

Having raised his hopes about a potential new home, Participant 9 found it difficult to understand that an accommodation supervisor who did not know him or his circumstances could override the decision of the police involved in his case and take this hope away at the last moment. This extract illustrates the extent of the lack of control that he had over his situation, even over a decision as personal as where he would live, the impact this had on his life, and what he was able to do. Participant 9 was trying to be patient with this situation, but was finding it frustrating.

The types of consequences of registration and restrictions cited by participants were very similar to those identified in previous research, such as Levenson and Tewksbury (2009) who also found an impact on employment and housing options, depression, hopelessness, and fear for personal safety. These types of hurdles would be closest to what Burchfield and Mingus (2008) described as formal barriers to social capital networks.

> "I always quote Timpsons…who do so much work with ex-offenders, working with offenders in prison…which is brilliant but won't touch a sex offender…and I'm feeling, you know well why? Everybody has, I guess their own, it's up to them…but to me, that's discrimination." (Ptpt 9, extract 25)

Participant 9 found it frustrating that even employers who ostensibly support 'ex-offenders' do not include those convicted of sexual offences, and he felt discriminated against by employers twice as a result, once for having a conviction, and then again by more inclusive employers, for having a sexual offence conviction. This is an example of 'intersectional' stigma (Turan et al., 2019); Participant 9 had two stigmatized identities, which resulted in additional negative perceptions and treatment. Discrimination usually refers to unfair treatment because of a protected characteristic, i.e. who someone is or something they did not choose, so it is interesting that Participant 9 used this term in relation to his offending history. Perhaps he felt his sexual interest in children was also something he did not choose, although he had chosen to act on it.

Participant 2 focused on how he felt being on the register meant he could get shamed at any time, that this was an indignity, and that a sexual offence was the only offence where someone could still be punished after serving their sentence:

> "The worst thing is your indignity…it means you have to go to the police station once a year… er the police officer can come any time to your house…er they can arrest you for no reason…obviously they got their own back-up, they can say yep you breached… Any time, it doesn't matter if your time even runs out, even you spent your time, it's the only conviction you can do time for, the time, and you still do more time." (Ptpt 2, extract 26)

Extract 26 also appears to be related to control; whatever Participant 2 did, the police could arrest him or return him to prison. He did not see this as being dependent on his behaviour so could never relax or let his guard down, he would always be 'do(ing) more time' and could have his freedom taken away again at any time. He too was aware that the restrictions on him were directly related to his offence having been sexual in nature, which had led to a feeling of helplessness about anything being able to change in the future, and a lack of control over his own life.

Participant 8 went further in how he described his current situation:

> "I would say that I live a half-life now…yeah…I don't do anything really you know I don't do a lot of things, I go to the gym and that's it, my life is very lonely." (Ptpt 8, extract 27)

It is interesting to consider what a 'half-life' meant to Participant 8. He saw his life as diminished in some way; he was very limited in the activities he could do and the relationships he could have, and did not feel he was living his life to its full potential. He had also previously reported attending a community project for MCOSOs, meeting with Probation, and even going on dates. However, he only mentioned going to the gym when considering what his life consisted of; he may have not included the other activities because he was trying to illustrate how stark his life now seemed, or perhaps he already saw these other activities as directly impacted on by his offences, whereas the gym may have been the one place he could just be another member of the public. A 'half-life' also indicates that Participant 8 did not feel fully alive at this time and had blocked off much of his emotional experience until he could come off the 'sex offender' register.

It is also of note that Participant 8 stated 'my life is very lonely' as opposed to 'I am very lonely' or 'I feel very lonely'. The latter options are more emotive to speak out loud, so referring to his life rather than himself allowed some distance from the sentiment while still sharing it with the researcher. He may also find it easier to separate himself and his feelings from his life generally, in terms of being able to cope with this daily.

There's no Way to Build up Trust

Even when the participants had been out of prison for a prolonged period of time and had not given any cause for concern, they felt there was no recognition of this development of trust or demonstration of managing their risks effectively; the stigma attached to them remained the same.

> "...obviously they're very helpful in the fact that you know they're not just there to be awkward. You know I don't believe that for one minute, you know they're there for a reason...erm, I don't always know that reason... it's about protecting me as much as others, you know it's not just about protecting the public you know it's protecting me... erm so I do understand that, but like I say, er as time goes on you know you've got you know it's it's up to me to earn that trust again, and then they're telling you how well you're doing and how you know you've progressed and everything. But then you ask for something, you know and it's 'no'. They keep telling me how well I'm doing and how forward I've changed, you know things have moved on... but I don't see any evidence from their side where they're trusting me any more." (Ptpt 9, extract 28)

Participant 9 was receiving support and praise for progress made, but had not been given the opportunity to prove he had moved on, and professionals' trust in him had not increased over time. Hope has been identified as a key part of the process for change (Weaver, 2014), and Participant 9 appeared to have little hope. Loftland (1969, cited in Maruna, 2001) emphasised the importance of 'normal-smiths', i.e.

people who can see 'ex-offenders' as 'normal' and recognise that they have been doing well. This can be significant for people who do not expect society to recognise the progress they have made. For Participant 9, this element was missing to a certain extent; he received verbal praise that he was doing well, but no corresponding changes in how he was managed. Caspi (1993, cited in Maruna, 2001, p.366) highlighted how performance accomplishments can increase a 'person's sense of self-efficacy and appraisal of internal control'; Participant 9 was not receiving much recognition of the positive goals he had achieved or his desistance and the focus was still more on things he was not allowed to do to prevent him re-offending.

> "I'm a drug addict I can say well I had loads of drugs, I was arrested and I spent five years in prison cos I had those drugs, now I'm three years clean… what am I going to say to you? If I'm never going to be a rapist again, do you know what I mean.
>
> Interviewer: Yeah, do you mean like you can't prove it in the same way?
>
> Ptpt: Yes, there's no like you know I can't say to you I've no interest in raping children… And you know that that's the truth… Whereas if I said here's three years of drug tests…Do you know what I mean? So it's hard to be able to put into context what you're explaining." (Ptpt 8, extract 29)

It is true that desistance from sexual offending can never be 'proved' other than through time; the message being communicated to 'ex-offenders' may therefore be that others believe they could reoffend at any time and will need monitoring indefinitely, indeed some will be on the 'sex offender' register for life. Loftland (1969, cited in Maruna, 2001, p.210) stated that 'even outstanding conformity is likely always to be greeted by (…) suspicion and fear'. This is certainly the message sent by media coverage of sexual offences, and the general public also tends to believe people who have committed sexual offences reoffend at high rates (Levenson et al., 2007). In fact, the base rate for sexual reoffending is relatively low, with a meta-analysis by Hanson and Morton-Bourgon (2006) finding a recidivism rate of 13.7%. Knowing that the public considers MCOSOs to be likely to reoffend, despite evidence to the contrary, could make it difficult to move forward or to consider oneself truly to be desisting.

Restrictions Not Fit for Purpose

Restrictions not being individual enough was a common thread amongst participants and referred to individual circumstances, motivations and offences not being taken into account.

Participant 1 described his restrictions as:

> "Too many, they're too cold, too blind, erm, at the end of the day, they're not very individual." (Ptpt 1, extract 30)

Participant 1's words indicate he saw his restrictions as excessive and harsh, that they were set without taking the whole picture into account, and that not enough attention was paid to how individuals on the register feel.

> "They've got to do checks for like underage kids that are living there and I think that's like unfair. With me anyway like I haven't committed no offence against kids or nothing." (Ptpt 3, extract 31)

Participant 3 needed permission to visit a relative with children, despite his offence having been committed against an adult, and there being no indication of him posing any risk towards children. This suggests a potential misunderstanding of his offending and risk factors, which meant not being able to have contact with his family, as well as the stigma of being considered risky to his nieces/nephews, with no evidence to support this. Weaver (2011) proposed involving service users in decisions around their management; ensuring restrictions are more personalised could also help in developing agency, which has been linked to desistance (Farmer et al., 2012). Licence conditions must be proportionate to a person's risk, but participants saw some of their restrictions as less justified. The literature is mixed, however, on whether it is reasonable to impose restrictions related to other potential victims. The percentage of people found to have offended against both child and adult victims has ranged from 8% (Cann et al., 2007) to 48% (Sim & Proeve, 2010); Kleban et al. (2013) found a crossover between age of victims of 5% when excluding adolescent victims (40% when they were included).

Another example around restrictions not being fit for purpose concerned who participants could associate with.

> "The SOPO, you've got loads of restrictions on it and that's what causes a lot of the problems... You know like if you've got friends, you've known for years, and all of a sudden you can't talk to them because you've just found out they're on the register... that's really hard to go through." (Ptpt 7, extract 32)

Interestingly, Participant 7 used the word 'sudden' regarding finding out that long-term friends of his were also on the 'sex offender' register. This could have been a hypothetical statement, however, he said 'that's really hard to go through' suggesting he had experienced it. He was still expressing sexual thoughts about children, so it may not have been beneficial for him to associate with others who could perhaps normalise these thoughts. However, MCOSOs tend to have restrictions on associating with others with sexual offence convictions. This is not necessarily based on individual risk assessment, and exists despite large numbers of MCOSOs having offended alone and there being little evidence that having a friend with a similar conviction would increase their risk of reoffending. It could be argued that, for some, social support from a peer (as in the 12-step approach) could be more helpful than staying away from other 'ex-offenders' and being socially isolated, although there is a need for research in this area.

Furthermore, if restrictions prevent people from associating with those who may be more accepting of their past, as well as making it harder to meet or be accepted by new people, who is left for them to get support from? This is important considering the evidence that being able to create pro-social networks and intimate relationships can help reduce recidivism (Willis et al., 2010), while social isolation increases the likelihood of reoffending (Kruttschnitt et al., 2000).

Some participants raised the issue of fairness surrounding their restrictions, how these were applied, or employment opportunities. While some acknowledged that others may not agree with this perception, due to their offending, there may still be a mismatch between what a member of the general public and the participants would consider unfair in the situations cited here.

These could contribute to the wider discussion around the role of punishment as opposed to rehabilitation or public protection. The Ministry of Justice Green Paper (2010) stated that 'criminals should face the robust and demanding punishments which the public expects. There must be consequences for breaking the law' (p.9). Some restrictions on MCOSOs, which do not appear to reduce reoffending or aid reintegration, may more accurately be described as punishments, or 'residual punishment(s)' (Hamilton, 2017). Barnett and Fitzalan Howard's (2018) review of meta-analyses and systematic reviews exploring ineffective strategies in reducing reoffending found that punishment-based interventions are less likely to reduce reoffending and, without rehabilitative support, can actually increase it. People with convictions concerned by these measures need to understand why a particular punishment is being received and to perceive it as fair (Fitzalan Howard & Wakeling, 2020). While participants did not express any plans to ignore their restrictions and feared being returned to prison, 'ex-offenders' can be less likely to comply with laws they view as unfair (Makkai & Braithwaite, 1994; Sherman, 1993). It is important, therefore, that those on the 'sex offender' register see their restrictions as relevant to them and their offending, rather than a blanket rule which does not impact on their individual risks.

Negative Psychological Impact

As well as the physical restrictions on behaviour, participants talked about the psychological implications of knowing they are on a 'sex offender' register and the difficulties of coping with the associated feelings.

> "Interviewer: How do you feel knowing that you're on the register?
>
> Ptpt: I mean it's heart-breaking isn't it, it crushes you inside…it kills me inside, yeah, if I'm honest, I can't really cope with it, I find it really difficult. You know I find it, that it's like a continuous reminder, reminder of punishment, of the incident that I've done… things I've done and I know a lot of that's probably me internally, you know it's not necessarily that, it's what the pressure I put on myself…you know how I feel about my offending you know… being on the register, I mean to me it just frustrates me that I can't do loads of things I want to do, do you know what I mean… that's the main main thing you know… and it's frustrating, it's life limiting, you know and I understand why because I've limited somebody else's life do you know what I mean. But I just feel that it shouldn't be continuous punishment." (Ptpt 8, extract 33)

Participant 8 used extremely emotive metaphors to describe his feelings about being on the register: 'heart-breaking', 'crushes you inside', 'kills me inside', and the tense he used indicated these feelings were ongoing. He described them as being more relevant to his remorse, and the register being a constant reminder of that, but the feelings he referenced all related to his life ending. Once he had offended, he saw his life as effectively over. He used this metaphor to describe how he felt psychologically, that what he had hoped his life would be was no longer possible. He felt unable to cope with the feelings of acknowledging what he had done and felt 'crushed' as a result. He was making efforts to move away from a past self, as he found it difficult to accept his actions, and perhaps in doing so, also trying to avoid a 'feared future self' (Paternoster & Bushway, 2009) or at least others perceiving him as this feared self. In this way, he was attempting to transform his identity, which can be important to the desistance process (Maruna & LeBel, 2009).

However, the idea of continuous punishment is in opposition to the possibility for redemption narratives (Maruna, 2001), and measures which are purely punitive are likely to damage social relationships, rather than promote desistance (Weaver & McNeill, 2015). Participant 8 also referred to being on the register as 'life-limiting'; he understood that he had limited his victim's life, but did not feel his punishment should be forever. He was experiencing public stigma in all the ways described by Herek (2007, 2009): enacted stigma (being treated negatively), felt stigma (experiencing and anticipating stigmatization) and internalised stigma (feeling less worthy and being distressed by this).

Participants did not generally express worries about reoffending, but did worry about being suspected or accused of another offence, and word 'getting out' to friends and neighbours. This was very difficult to cope with, given that others' perceptions and actions were out of their control. This led to them using various ways to attempt to manage these concerns, including having to consider the potential consequences of starting a new relationship:

> "I think it stops me from (sighs) sort of er pursuing friendships more...erm, like I say I don't mind general chit chat…but getting you know closer to someone." (Ptpt 9, extract 34)

Participant 9 worried about people finding out about his offending and so avoided getting close to people; his relationships were mostly superficial, and he did not allow himself to develop new, deeper friendships. This worry and subsequent avoidance served as both an individual and community barrier (Burchfield & Mingus, 2008) to him seeking and having support, and to integrating into the community through these relationships. As above, Willis et al. (2010) highlighted that being able to create pro-social networks and intimate relationships can lead to a reduction in recidivism.

Discussion

This chapter has highlighted the idiographic perspectives of men on the 'sex offender' register, and it is evident that being on the register and bearing the label of 'sex offender' is extremely difficult both practically and psychologically. Individuals attempting to reintegrate into the community may need more support, specifically to manage their new identity. MCOSOs will largely already be aware of how they are viewed. It is important to allow space to process the challenges and stigma associated with having a sexual offence conviction (e.g. in treatment programmes or supervision with Probation) and consider how to balance taking responsibility for their actions without allowing their offence to become their master status (Becker, 2018). For some, this would include access to support with addressing their own traumatic experiences and understanding how these impacted on their decision-making to offend. Supporting individuals to develop their personal strengths and protective factors would also help in allowing them a balanced view of themselves as a whole person.

Continuing to increase awareness and empathy of the difficulties faced by MCOSOs in reintegrating, as well as their potential impact, could support developing more effective working relationships, particularly with police officers overseeing those on the register. Considering terminology, for example, moving away from using the term 'sex offender' in favour of MCOSO is also a small action but which can have a big impact.

Key areas for policymakers to consider include how individual restrictions are set and reviewed, particularly the impact on their ability to access housing, employment and build pro-social networks, all of which can impact desistance (Willis et al., 2010). While professionals may understandably be inclined to be risk-averse due to concerns about the implications of further offending, being mindful of reconviction risk data for this population is important so that people are treated as individuals and plans implemented according to their individual risks and needs, rather than their label. This could also prevent people from continuing to see themselves as inherently dangerous and undeserving of redemption (Ward & Maruna, 2007).

Wider questions arise over whether the 'sex offender' register is fit for purpose, and if alternative strategies could better support individuals, whilst also promoting reintegration and desistance. This research also raises questions about the possible impact of the challenges men on the 'sex offender' register face on their ability or motivation to desist. This study did not consider reconviction, but previous research has highlighted a potential link between not being able to reintegrate into the community, and risk of reoffending (e.g. Tewksbury, 2007) as well as a lack of evidence for such restrictions reducing offending (e.g. Tewksbury, 2007; Bowen et al., 2016). If restrictions are not having the desired effect, and are in fact having a detrimental impact, which in turn may further reduce their effectiveness, it is important to consider whether the public favouring such policies (e.g. Levenson et al., 2007) warrants continuing to use them in their current form. Applying more specific restrictions less indiscriminately, whilst also prioritising supporting MCOSOs in their resettlement and desistance, could benefit not only these individuals but also any potential further victims.

One of the aims of this study was to understand the idiographic perspectives of men on the 'sex offender' register and what their experiences have meant for them and their identities. When negotiating their new identity as a 'sex offender', participants were very aware of negative perceptions by others, and found this difficult to manage. Some had reconciled their offending with their new identities, and 'made good' (Maruna, 2001); some no longer saw themselves as a 'sex offender' and had detached themselves from this identity (as in Crocker & Major, 1989; Miller & Kaiser, 2001) but even so they struggled to maintain this view in the face of public opinion. Participants also spoke about being fearful of others finding out about their offending and registration status; some feared this may lead to physical reprisals or had experienced this already. They were also afraid of being falsely accused of doing something wrong or committing another offence. This theme adds to the picture of participants' sentences or punishment not yet being over, and perhaps never being able to feel that it is over. This leaves questions about the message that is communicated to those who have committed sexual offences, and whether they can ever move on from this.

The superordinate theme of 'Negative impacts of the register and restrictions' highlighted the number of difficulties faced by participants, such as restrictions not being fit for purpose and preventing individuals from progressing. This fits with previous research, such as Levenson (2008), who found that being on a register impacted employment and housing, and Burchfield and Mingus (2008), who described barriers to social capital networks for those on a register. There was also a clear negative psychological impact on participants, including constant worry about their lives 'falling apart again', and inability to be honest with those around them. This fits with previous research, such as Hatzenbuehler (2009), who found that when people experience stigma-related stressors, they report more psychological distress. Participants attempted to live with the restrictions as best they could. Some focused on gaining purpose from, for example, employment or generative activities (Maruna, 2001); others avoided developing relationships and going anywhere they thought might put them at risk, much as in Bos et al.'s (2013) work. These were also in line with some of the approaches identified by Evans and Cubellis (2015) as being used by 'registered sex offenders' to manage stigma, particularly honesty, concealment and isolation, but also denial for Participant 2. Grouping was not easy for participants to achieve due to restrictions. Participants appreciated the support they had, whether this was from their Probation Officer, or those who still had support from some family or friends. However, they advocated for needing more support. More needs to be done to provide accessible, meaningful support for those who are limited in the services they can access.

Conclusion

This study focused on the experiences of reintegration and being on the 'register' for men with sexual convictions. The study found that being on the register and bearing the label of 'sex offender' is extremely difficult both practically and psychologically. Being on the 'sex offender' register was reported to have fewer practical implications than other licence restrictions or orders in place, however, even the awareness of being on the register had a significant impact on their view of themselves and others. There are daily challenges in managing relationships, employment and accommodation, as well as longer-term challenges such as reconciling a past offence with a current identity. This data raises questions about the possible impact of these challenges on desistance or reoffending. While clearly this study did not consider reconviction or seek to verify information from participants about their offending, previous research has highlighted a potential link between not being able to reintegrate into the community and risk of reoffending (e.g. Tewksbury, 2007).

References

Barnett, G.D., & Fitzalan Howard, F. (2018). What doesn't work to reduce reoffending? A review of reviews of ineffective interventions for adults convicted of crimes. European Psychologist, 23(2), 111–129. doi:https://doi.org/10.1027/1016-9040/a000323

Becker, H. (2018). *Outsiders*. Free Press. Original work published 1963.

Bowen, K., Frenzel, E., & Spraitz, J. D. (2016). Sex offender registration and notification laws: Thoughts from registered sex offenders in three states. *Safer communities, 15*(2), 94–109. https://doi.org/10.1108/SC-12-2015-0040

Bows, H., & Westmarland, N. (2018). Older sex offenders—managing risk in the community from a policing perspective. *Policing and Society, 28*(1), 62–74. https://doi.org/10.1080/10439463.2016.1138476

Brown, K., Spencer, J., & Deakin, J. (2007). The reintegration of sex offenders: Barriers and opportunities for employment. *The Howard Journal of Criminal Justice, 46*(1), 33–42. https://doi.org/10.1111/j.14682311.2007.00452.x

Burchfield, K. B., & Mingus, W. (2008). Not in my Neighborhood: Assessing registered sex offenders experiences with local social capital and social control. *Criminal Justice and Behavior, 35*(3), 356–374. https://doi.org/10.1177/0093854807311375

Cann, J., Friendship, C., & Gonza, L. (2007). Assessing crossover in a sample of sexual offenders with multiple victims. *Legal and Criminological Psychology, 12*, 149–163. https://doi.org/10.1348/135532506X112439

Crocker, J., & Major, B. (1989). Social stigma and self-esteem: The self-protective properties of stigma. *Psychological Review, 96*(4), 608–630. https://doi.org/apa.org/psycinfo/1990-04498-001

Cubellis, M. A., Evans, D. N., & Fera, A. G. (2019). Sex offender stigma: An exploration of vigilantism against sex offenders, *Deviant Behavior, 40*(2), 225–239. https://doi.org/10.1080/01639625.2017.1420459

De Wet, J., & Erasmus, Z. (2005). Towards rigour in qualitative analysis. *Qualitative Research Journal, 5*(1), 27.

Evans, D., & Cubellis, M. (2015). Coping with stigma: How registered sex offenders manage their public identities. *American Journal of Criminal Justice, 40*(3), 593–619. https://doi.org/10.1007/s12103-014-9277-z

Farmer, M., Beech, A. R., & Ward, T. (2012). Assessing desistance in child molesters: A qualitative analysis. *Journal of Interpersonal Violence, 27*(5), 930–950. https://doi.org/10.0.4.153/0886260511423255

Fitzalan Howard, F., & Wakeling, H. (2020). People in prisons' perceptions of procedural justice in England and Wales. *Criminal Justice and Behavior, 47*(12), 1654–1676. https://doi.org/10.1177/0093854820964501

Grossi, L. M. (2017). Sexual offenders, violent offenders, and community reentry: Challenges and treatment considerations. *Aggression and Violent Behavior, 34*, 59–67. https://doi.org/10.1016/j.avb.2017.04.005

Hamilton, E. (2017). Identity concerns among sexual offenders: The narrative call. *Practice Innovations, 2*(1), 13–20. https://doi.org/10.1037/pri0000039

Hanson, R. K., & Morton-Bourgon, K. E. (2006). Characteristics of persistent sexual offenders: A meta-analysis of recidivism studies. *Journal of Consulting and Clinical Psychology, 73*(6), 1154–1163. https://doi.org/10.1037/0022-006x.73.6.1154

Harris, D. A. (2014). Desistance from sexual offending: Findings from 21 life history narratives. *Journal of Interpersonal Violence, 29*(9), 1554–1578. https://doi.org/10.1177/0886260513511532

Harris, M., Tynan, R., & Edgar, K. (2020). Thinking differently: Employers views on hiring people convicted of sexual offences. Prison Reform Trust.

Hatzenbuehler, M. L. (2009). How does sexual minority stigma "get under the skin"? A psychological mediation framework. *Psychological Bulletin, 135*(5), 707–730. https://doi.org/10.1037/a0016441

Herek, G. M. (2007). Confronting sexual stigma and prejudice: Theory and practice. *Journal of Social Issues, 63*, 905–925. https://doi.org/10.1111/j.1540-4560.2007.00544.x

Herek, G. M. (2009). Sexual stigma and sexual prejudice in the United States: A conceptual framework. In D. A. Hope (Eds.), *Contemporary perspectives on lesbian, gay and bisexual identities: The 54th Nebraska Symposium on Motivation* (pp. 65–111). Springer.

Hochstetler, A., Copes, H., & Williams, J. P. (2010). Thats not who I am: How offenders commit violent acts and reject authentically violent selves. *Justice Quarterly, 27*(4), 492–516. https://doi.org/10.1080/07418820903173344

Kemshall, H. (2008). *Understanding the community management of high risk offenders*. McGraw Hill/Open University Press.

Kemshall, H., & Maguire, M. (2001). Public protection, partnership and risk penality: The multi-agency risk management of sexual and violent offenders. *Punishment & Society, 3*(2), 237–264. https://doi.org/10.1177/14624740122228311

Kewley, S., & Brereton, S. (2022). Public protection: Examining the impact of strengthened public protection policy on probation practice. In L. Burke,

N. Carr, E. Cluley, S. Collett and F. McNeill (Eds.), *Reimagining probation practice* (pp. 112–131). Routledge.

Kleban, H., Chesin, M. S., Jeglic, E. L., & Mercado, C. C. (2013). An exploration of crossover sexual offending. *Sexual Abuse: A Journal of Research and Treatment, 25*(5), 427–443. https://doi.org/10.1177%2F1079063212464397

Kruttschnitt, C., Uggen, C., & Shelton, K. (2000). Predictors of desistance among sex offenders: The interaction of formal and informal social controls. *Justice Quarterly, 17*, 61–87. https://doi.org/10.1080/07418820000094481

Lasher, M. P., & McGrath, R. J. (2012). The impact of community notification on sex offender reintegration: A quantitative review of the research literature. *International Journal of Offender Therapy and Comparative Criminology, 56*(1), 6–28. https://doi.org/10.1177/0306624X10387524

Lester, J. (2008). Brandishing the Mark of Cain. Federal Sentencing Reporter. http://fsr.ucpress.edu/content/ucpfsr/21/2/107.full.pdf

Levenson, J. S. (2008). Collateral consequences of sex offender residence restrictions. *Criminal Justice Studies, 21*(2), 153–166. https://psycnet.apa.org/doi/10.1080/14786010802159822

Levenson, J. S., Brannon, Y. N., Fortney, T., & Baker, J. (2007). Public perceptions about sex offenders and community protection policies. *Analyses of Social Issues & Public Policy, 7*(1), 137–161. https://doi.org/10.1111/j.1530-2415.2007.00119.x

Levenson, J., & Tewksbury, R. (2009). Collateral damage: Family members of registered sex offenders. *American Journal of Criminal Justice, 34*(1–2), 54–68. https://doi.org/10.1007/s12103-008-9055-x

Makkai, T., & Braithwaite, J. (1994). Reintegrative shaming and compliance with regulatory standards. *Criminology, 32*, 361–385. https://doi.org/10.1111/j.1745-9125.1994.tb01158.x

Markus, H., & Nurius, P. (1986). Possible selves. *American Psychologist, 41*(9), 954–969. https://doi.org/10.1037/0003-66X.41.9.954

Maruna, S. (2001). Making good: How ex-convicts reform and rebuild their lives. *American Psychological Association.* https://doi.org/10.1037/10430-000

Maruna, S., & LeBel, T. P. (2009). Strengths-based approaches to reentry: Extra mileage toward reintegration and destigmatization. *Japanese Journal of Sociological Criminology, 34*, 59–80. https://doi.org/10.20621/jjscrim.34.0_59

Mears, D. P., & Cochran, J. C. (2015). *Prisoner Reentry in the Era of Mass Incarceration.* Sage Publications.

Miller, C.T., & Kaiser, C. R. (2001). A theoretical perspective on coping with stigma. *Journal of Social Issues, 57*, 73–92. https://psycnet.apa.org/doi/10.1111/0022-4537.00202

Ministry of Justice. (2010). Breaking the Cycle: Effective Punishment, Rehabilitation and Sentencing of Offenders. Presented to Parliament by the Lord Chancellor and Secretary of State for Justice by Command of Her Majesty.

Ministry of Justice (2025). Offender Management Statistics Bulletin, 30 October 2025.

Paternoster, R., & Bushway, S. (2009). Desistance and the feared self: Toward an identity theory of criminal desistance. *The Journal of Criminal Law and Criminology, 99*(4), 1103–1156. http://www.jstor.org/stable/20685067

Rydberg, J. (2018). Employment and Housing Challenges Experienced by Sex Offenders during Reentry on Parole. *Corrections, 3*(1), 15–37. https://doi.org/10.1080/23774657.2017.1369373

Seidler, K. (2010). Community management of sex offenders: Stigma versus support. *Sexual Abuse in Australia and New Zealand*. http://search.proquest.com/openview/83eb15aa0fdc84329616abae59d2a3e2/1?pq-origsite=gscholar

Sentencing Council. (2014). Sexual offences—definitive guideline.

Sherman, L. W. (1993). Defiance, deterrence, and irrelevance: A theory of the criminal sanction. *Journal of Research in Crime and Delinquency, 30*, 445–473. https://doi.org/10.1177%2F0022427893030004006

Sim, D. J., & Proeve, M. (2010). Crossover and stability of victim type in child molesters. *Legal and Criminological Psychology, 15*, 401–413. https://doi.org/10.1348/135532509X473869

Smith, J. A. (2015). Qualitative psychology: A practical guide to research methods. Sage Publications.

Smith, J. A., & Osborn, M. (2003). *Interpretative phenomenological analysis*. In J. A. Smith (Ed.), Qualitative psychology. A practical guide to research methods (pp. 51–80). Sage Publications.

Smith, J., & Eatough, V. (2007). *Interpretative phenomenological analysis*. In E. Lyons, & A. Coyle (Eds.), Analysing qualitative data in psychology (pp. 35–50). Sage Publications.

Tewksbury, R. (2007). Exile at Home: The Unintended Collateral Consequences of Sex Offender Residency Restrictions. *Harvard Civil Rights-Civil Liberties Law Review*, 42, 531–540. http://ilvoices.org/uploads/3/4/1/6/34164648/70-tewksbury-summer2007.pdf

Tewksbury, R., & Lees, M. (2006). Perceptions of sex offender registration: Collateral consequences and community experiences. *Sociological Spectrum, 26*(3), 309–334. https://doi.org/10.1080/02732170500524246
Tolson, D., & Klein, J. (2015). Registration, residency restrictions, and community notification: A social capital perspective on the isolation of registered sex offenders in our communities. *Journal of Human Behavior in the Social Environment, 25*(5), 375–390. https://doi.org/10.1080/10911359.2014.966221
Tovey, L., Winder, B., & Blagden, N. (2022). '"It's ok if you were in for robbery or murder, but sex offending, that's a no no": A qualitative analysis of the experiences of 12 men with sexual convictions seeking employment'. *Psychology, Crime and Law, 29*(6), 653–676. https://doi.org/10.1080/1068316X.2022.2030736
Turan, J. M., Elafros, M. A., Logie, C. H., Banik, S., Turan, B., Crockett, K. B., Pescosolido, B., & Murray, S. M. (2019). Challenges and opportunities in examining and addressing intersectional stigma and health. *BMC Medicine, 17*(7). https://doi.org/10.1186/s12916-018-1246-9
Ward, T., & Maruna, S. (2007). *Rehabilitation: Beyond the risk paradigm.* Routledge.
Weaver, B. (2011). Co-producing community justice: The transformative potential of personalisation for penal sanctions. *British Journal of Social Work, 41*, 1038–1057. https://doi.org/10.1093/bjsw/bcr011
Weaver, B. (2014). Control or change? Developing dialogues between desistance research and public protection practices. *Probation Journal, 61*(1), 8–26. http://prb.sagepub.com/content/61/1/8.short
Weaver, B., & McNeill, F. (2015). Lifelines: Desistance, social relations and reciprocity. *Criminal Justice and Behavior, 42*(1), 95–107. https://doi.org/10.1177/0093854814550031
Willis, G. M., Levenson, J. S., & Ward, T. (2010). Desistance and attitudes towards sex offenders: Facilitation or hindrance? *Journal of Family Violence, 25*(6), 545–556. doi:https://doi.org/10.1007/s10896-010-9314-8
Winder, B., & Gough, B. (2010). I never touched anybody—Thats my defence: A qualitative analysis of internet sex offender accounts. *Journal of Sexual Aggression, 16*(2), 125–141. https://doi.org/10.1080/13552600903503383
Winnick, T. A., & Bodkin, M. (2009). Stigma, secrecy and race: An empirical examination of black and white incarcerated men. *American Journal of Criminal Justice, 34*(1–2), 131–150. https://doi.org/10.1007/s12103-008-9050-2

4

Navigating Life with a Sexual Conviction: 'Anticipatory Disclosure' and the Impact of Living in a Perpetual Biopsychological State of Fight or Flight

Kirsty Teague

Introduction

The "pains" of having a sexual conviction have been cited by Ievins (2022, 2023) in their work on the society of 'sex offenders' and the stains of imprisonment. Indeed, much work exists analysing the situated experiences of men navigating life with a sexual conviction in prison (Blagden et al., 2019; Ievins & Crewe, 2015) and to an extent, in the community (McCartan & Richards, 2021; Saunders, 2022; Teague, 2022). Whilst such literature is critical for understanding some of the broader lived realities of those who have a sexual conviction, there is a paucity of understanding on the nature and impact of more nuanced aspects of living with a criminal conviction for a sexual crime. One such area in particular relates to 'disclosure', the process of disclosing criminal record history to another individual, organisation, or in a particular forum. Interestingly, the notion and process of disclosure in regard to

K. Teague (✉)
University of Derby, Derby, UK
e-mail: k.teague@derby.ac.uk

N. Blagden et al. (eds.), *Sexual Crime and Community Reintegration*, Sexual Crime,
https://doi.org/10.1007/978-3-032-06034-1_4

individuals who have a sexual conviction has not been subject to critical appraisal encompassing the totality of what it entails, to whom, and in what context. Instead, much information and literature around disclosure centres around necessary, but practically oriented, guides relating to when, and how, to disclose in the process of applying for jobs, college, or university (see for example: Nacro, 2023; Unlock, 2023). This chapter utilises psychosocial theory to critique (i) the socio-legal landscape surrounding disclosure; (ii) the period before disclosing—'anticipatory disclosure'; and (iii) the impacts of disclosing a conviction for a sexual offence. To do this, the chapter sets out the realities of resettlement before conceptualising 'anticipatory disclosure'. Following this, the chapter subsequently develops a critical discussion around the implications emergent from such conceptualisation, and recommendations for policy and practice.

Realities of Resettlement

For many serving a custodial sentence, the (often) long-awaited anticipation of leaving prison is matched with apprehension for what life post-imprisonment 'looks like'. Western et al. (2015, p. 1512), for example, found that "anxiety and feelings of isolation accompanied extreme material insecurity". Exacerbating this, having a conviction for a sexual crime can further intensify concern due to the correlation with stigma and vigilantism crimes (Cubellis et al., 2019). This section will adopt a micro, meso and macro approach (Serpa & Ferreira, 2019) to analysing the realities of resettlement for men who have a sexual conviction. Permeating all levels of social analysis is the primacy of risk (perception). The rise of possibilistic risk, and of the precautionary culture, creates a tension between possibilistic and probabilistic risk perception and assessment (Furedi, 2009). The surgency behind the rise, and influence, of possibilistic thinking has been "driven by a powerful sense of cultural pessimism about knowing and an intense feeling of apprehension about the unknown. The cumulative outcome of this sensibility is the routinisation of the expectation of worst possible outcomes" (ibid, pg. 205). The many consequences of this centre around impediments in protections relating to human rights and civil liberties.

On a macro level, the narratives perpetuated by politicians and media outlets inform public understandings around risk and what is deemed to be a cause for concern. As such, the authority behind the narrative can be regarded as legitimate and thus unquestionable (Tankebe & Liebling, 2014). This coalesces with a lack of public education around matters of crime and justice which allows misinformation to circulate, and consequently fear and moral panics to emerge (Pickett et al., 2015). Scholars publishing in the realm of the sociology of vindictiveness note the impact of censure and vindictiveness in contributing to othering, and the consequential imposition of potentially harmful, binarised ways of thinking (Mahoney et al., 2022; Teague & Winder, 2023). This ties in with understandings around stigma, and stigmatised identities (see Goffman, 1963; Tyler, 2020). Relatedly, Ievins (2022, 2023) conceptualises the result of stigma-based exclusion through the development of moral community to counter the pain and 'stain' of having a sexual conviction. As such, having a sexual conviction has been described as having 'social curse'-like qualities, which require meaningful and sustained social support to mitigate the effects of the negative consequences (Bell et al., 2018; Blagden et al., 2020). A significant issue emerges here upon considering the repercussions of founded issues in forming and maintaining meaningful social support.

Pro-social connection can be founded in ways that might relate to where one studies, works, or actively strives for sociality. However, structurally, on a meso-level, there are a number of barriers to integration influenced by populist discourses which seek to denigrate people associated with sexual offending (Mahoney et al., 2022). In particular, education, employment, and accommodation are notable areas of challenge for all with a criminal record (see Henley, 2022). However, whilst often overlooked, means and opportunities for relationship-building (such as the internet) are also spaces which pose challenges for those with a sexual conviction (Saunders, 2022; Teague et al., 2021). In this way, restrictions posed as part of the community/public protection model (Kemshall & Maguire, 2001), and thus within the purview of risk management, can prevent (re)integration. Internet access, or lack thereof, and restricted access to 'members-only' clubs and organisations become a challenge to those who are seeking to form connection. Worldwide-operating platform, Meetup, whose tagline is: "where interests become friends" (Meetup, 2023b), is a popular platform of choice for adults to form social connections with like-minded adults. However, in keeping with risk appraisal in neoliberal society, the following places people with a sexual conviction in a precarious position:

> Meetup may remove certain members when we become aware that the member is a convicted sex offender or has been convicted of a violent, fraudulent, or dangerous crime. Meetup may also remove a member based on their criminal record, for safety, or for other concern (Meetup, 2023a)

As such, it is unsurprising that forming social connections can appear to be a quest marred with precarity and uncertainty. This is problematic in and of itself. However, it is especially problematic when considering the role that social isolation can have in increasing the risk of recidivism among individuals with a sexual conviction (Fox, 2017). This will be discussed further in the section to follow.

As recognised by Giddens (1997, p. 883), micro-sociology addresses the 'study of human behaviour in contexts of everyday direct interaction'; as outlined above, there are a number of structural ways in which the daily lives of men navigating life with a sexual conviction are challenged on a micro-level. Striving for a "good life" (Ward & Stewart, 2003) can be hampered by persistent hardships encountered consistently and routinely. Indeed, it must be noted also that, potentially aggravating such realities, are the impacts felt by non-offending family members, who often experience harsh treatment through association (Duncan et al., 2022; Armitage et al., 2023). As such, there is a transferability to the stigma firstly encountered by the individual associated with sexual offending (Cubellis et al., 2019). Consequently, the potential even for ripple effects can present challenges for forming connection. This is compounded by the legal and social implications of navigating life with a sexual conviction. Disclosing one's criminal record, herein referred to as disclosure, can be a barrier to forming meaningful connection for fear of having to disclose their criminal record.

The personal, social, and/or legal desire/need to disclose (a criminal record) can impact access to outlets for connection such as housing, employment, and thus also the potential for social and romantic connections to others, and the sustainability of maintaining such. Recently, Mahoney et al. (2022) have found that populist and vindictive responses to people associated with sexual offending are associated with pluralities of violence and can have negative implications for criminal and social justice. Relatedly, there is a connection to be drawn to the work of Goffman (1961) and his scholarly work on the mortification of self, and how an individual can go through a social death as a result of their negative experiences. Whilst the original conceptualisation related to forensic environments, the transferability of such a process is pertinent to a broader non-forensic environment, and thus insidiously applicable to a wider community setting.

> The new inpatient is cleanly stripped of many of his accustomed affirmations, satisfactions, and defences, and is subjected to a rather full set of mortifying experiences: restriction of free movement, communal living, diffuse authority of a whole echelon of people, and so on. Here one begins to learn about the limited extent to which a conception of oneself can be sustained when the usual supports for it are suddenly removed (Goffman, 1961, p. 148)

To mitigate the impacts of such, as discussed earlier, men with a sexual conviction may develop a moral community to counter the moral exclusion faced elsewhere in society (see Ievins, 2022, 2023). Shared challenges can offer the group, which is not homogenous by any means, opportunity to discuss, test, and challenge issues such as disclosure, and contribute to the building of a collective resilience. However, with "the volume of sexual offences recorded by the police [...] increasing over the last decade" (ONS, 2023, p. 2), and the limited opportunities for finding, or building, a community to navigate experience related to having a sexual conviction, this poses significant challenges to (i) individual wellbeing; (ii) the development of a civil society; (iii) (re)offending rates.

Understanding Anticipatory Disclosure in Context

Given the paucity of academic literature on the consequences of sexual offence criminal record disclosure, there is a lack of an understanding on the nature and extent of the impacts that the psychological and practical process has on the individual. More broadly, however, there has been literature published in areas relating to disclosure, particularly over the last two decades. Published literature here relates primarily, but not exclusively, to (i) privacy concerns (Pijoan, 2014; Weaver, 2018); (ii) technological developments and digital surveillance (Corda &Lageson, 2020; Lageson, 2022); (iii) "living in a shadow of the past" (Naylor et al., 2008; Thomas & Hebenton, 2013); and (iv) impacts on access, education and employment (Graber & Zitek, 2022; McTier, 2022; Naylor, 2005; Stewart & Uggen, 2020). Whilst there is a clear utility in this literature, there is a lack of work exploring how the period *prior* to disclosure is experienced. Conceptually, this can be referred to as 'anticipatory disclosure'.

Anticipatory disclosure here refers to the period psychologically and practically prior to disclosing one's criminal record. Psychologically, the experience of providing sensitive and personal information, brief or detailed, has the ability to cause significant distress. This might be likened experientially to perpetration-induced traumatic stress (PITS) (MacNair, 2002).

> PITS involves any portions of the symptomatology of PTSD, at clinical or subclinical levels, which result from situations that would be traumatic if someone were a victim, but situations for which the person in question was a causal participant (MacNair, 2002, p. 7)

Crucial here is the need to recognise heterogeneity in perpetration experiences, and that trauma might be linked to offending, and/or, potentially be the product of such. There is also a link to be drawn to the recognised overlap between those who have been harmed, and who have caused harm; this being "on the short list of criminological 'facts' [...] beginning with von Hentig's (1948) contention that perpetrators and victims are intimately linked" (Pyrooz et al., 2014, p. 316). In this way, it is important to recognise the often complex landscape that is present when considering the disclosure of a criminal record. Whilst it is imperative to flag the potential to cause secondary trauma (Figley, 1995), which may leave a trauma exposed unrecognised and/or unsupported, it may contribute to heighted risk of offending and otherwise harmful, or risky, behaviour. As such, it is important to recognise the importance, and role of, the environment/climate in which an individual discloses their past. There is much literature on the hostility surrounding people who have a criminal conviction, an environment crafted by moral entrepreneurs (rule creators) and moral crusaders (rule enforcers) who are concerned with the "creation and enforcement of deviant categories" (Becker, 1995, p. 169). The climate crafted thus can become fuelled by populism and leading to punitive punishment of those who have "deviated" from the "prototype of the rule creator" (ibid). This has been exemplified by the moral exclusion of people with a criminal record from various areas of social life by virtue of this aspect of their past. As such, it is unsurprising that the act of disclosing might be undesirable for the individual with the criminal record. This also needs considering against the backdrop of what individuals need to live a prosocial life, and desist from crime (see for example, Farmer et al., 2015; Hulley, 2016), and thus what is at stake when considering the unknown (i.e. the reaction) to disclosure. Parallel to this, literature around the "pains of release" highlight socially significant findings such as individuals reporting overstimulation, social disorientation, social precarity and missing "hooks" for new identities (McKendy & Ricciardelli, 2021). As discussed previously, men with a conviction for a sexual crime can feel a protracted "grip" of penal power (Crewe & Ievins, 2020), impacting their quality of social life in the community. Such context is critical in understanding how disclosure is experienced, or rather *felt*.

In discussing emotionality and feelings, it is imperative to consider how the period of anticipating disclosure can activate the "fight or flight response" (Cannon, 1929) in the individual, and question whether this is momentary, as the original conceptualisation puts forward, or whether it is more protracted, and perpetual, in the context of disclosure and what this can mean/lead to experientially. Such situation also needs positioning against the zemiological understanding of social harm. Zemiology here is:

> A word originating from Ancient Greece, zemia carries numerous connotations. Rather than simply meaning 'harm', as is so often intimated, the word actually denotes, among other things, loss, damage as well as various forms of punishment of deviant and/or legal transgressions (Allen, 2000; Boukli, 2017; Boukli & Kotzé, 2018, p. 2).

It could be suggested that the impacts, or harms if you will, of anticipating disclosure are "structurally induced harms" (ibid, p.4). This needs considering against the established link between having a conviction for a sexual crime and experiencing social anxiety, as founded by a metanalysis conducted by Nunes et al. (2012). For Porter et al. (2015, p. 44):

> "It is possible that the process of being convicted of a sex offense may promote the development of a psychiatric disorder through the impact of society's disdain and judgment, isolation from family and friends, being the lowest in the hierarchy of prison populations and often having been the victim of abuse" (Marshall et al., 2009)

As such, there is a founded need for the state, among other social actors, to *actively* support community reintegration. Ethically, there needs to be greater consideration of the impacts of the process of disclosure in its entirety, and its potential to exacerbate social anxiety among individuals with a sexual conviction. This is especially because social anxiety impacts the ability to (i) develop and maintain relationships whilst also sustaining social fears (Alden & Taylor, 2004) and (ii) is a risk factor for depression and substance misuse (Stein & Stein, 2008). There is thus a public health case to consider (see McCartan & Kemshall, 2023). This might be considered against the backdrop of there being a loneliness crisis (Holt-Lunstad, 2017, p. 129), requiring a public health approach in the way of social prescribing, which is a "a means of addressing mental, psychosocial, or socioeconomic issues, and enhancing community well-being and social inclusion" (Scottish Development Centre for Mental Health, in Chatterjee et al., 2018, p. 98). This recognition that an array of nonclinical interventions, such as social activities and social services, can help empower individuals and improve their health and well-being (Morse et al., 2022, p. 1) is one which has transferability to help support people with a criminal conviction, including for a sexual crime. However, there needs to be re-balancing of, often competing, requirements prior to this.

Whilst beyond the scope of this chapter to discuss the detailed and nuanced specifics around what the penal landscape for men with sexual convictions should look like in line with the evidence bases of 'what works' and desistance, it should be made clear that the current risk-centric approach to men with a sexual conviction in the community is both individually and socially harmful. Navigating a perpetual period of anticipating disclosure, in a climate which is vindictive and fuelled by populism and misinformation, is individually and socially harmful. This is not to say that disclosure itself is not necessary, or that it is always harmful. For some, as outlined in a case by Rennie and Crewe's (2023, p. 196), there can be a perceived utility in disclosing:

> He recognized that he was obliged to disclose his criminal history and was committed to doing so, orientating himself towards having to practice and get used to it, rather than seeing it as being inherently problematic.

However, there are two things to note here: (i) when presented with an 'obligation' that is persistent and life-limiting, this 'commitment' could be a positive reconstrual of something that would otherwise be negative (Blagden et al., 2020; McAdams, 2006); and (ii) there remains a founded problem with "having to practice and get used to it". As such, disclosure of a criminal record, for some, can be understood as an "invisible punishment" (Thomas & Hebenton, 2012, p. 238), which when layered with mass supervision, amounts to what McNeil refers to as "pervasive punishment" (McNeill, 2019). Here the penal landscape can be understood as more than just contributory to the mortification of self (Goffman, 1961), with a lack of acknowledgment of such an impact.

Implications, and Recommendations, for Policy and Practice

More recently, there has been published work critiquing existing policies and practices relating to men with sexual convictions and calls for the evidence base to inform and guide the curation of such (McCartan, 2022; Saunders, 2022). However, there has been a lack of literature adopting a public criminology approach to such. As a sub-discipline of criminology, public criminology (PC) advocates for going beyond mere acknowledgement, or understanding, of contemporary sociological and criminological issues to "make a difference and bring about change" (Carrabine et al., 2000, p. 207, emphasis added). In this way, PC clearly distinguishes itself from traditional ways of doing criminology: by transcending criminology's historically narrow boundaries, and through its commitment to actively avoiding the reproduction of "dominant constructions of 'crime' and justice" (Piché, 2015, p. 71). For public, and critical, criminologists, power is critical to understanding who defines what is criminal, and who is criminalised through the criminal justice system. To best understand drivers behind certain criminalised behaviours and other social harms, PC seeks to critique an array of socio-political issues, including homelessness, unemployment, health inequalities, and inadequate service provision. In this way, public criminologists actively concern themselves with understanding the status quo by working to improve services and conditions, whilst also actively engaging in local and national debates that lobby for policy development (Carrabine et al., 2000). Crucially, therefore, PC has at its core a concern with citizenship, social justice, and human rights (ibid). There are a number of PC principles and commitments which can offer a valuable theoretical lens through which to develop holistic-focused policy and practice, specifically: being (i) transparent; (ii) theoretically informed; (iii) evidence-based; (iv) empowerment-driven; (v) committed to practical change; (vi) committed to social justice and human rights; and (vii) committed to connecting public issues and private troubles (Carrabine et al., 2000). The need to address individuals' holistic needs is part of the drive for a "whole system" approach to people in the criminal justice system (Ministry of Justice, 2018). This section takes a micro, meso and macro approach (Serpa & Ferreira, 2019) to framing recommendations of change to inform policies and practices surrounding the lived realities of men who have a sexual conviction. Such recommendations will also be informed by the principles and commitments of public criminology.

At a macro level, for meaningful and sustained change there needs to be a clear government-supported strategy underpinned by research evidence on what works in rehabilitating and reintegrating men with sexual convictions. This needs to be focused on supporting the holistic needs of individuals and have at its core an advocation for a "whole system" approach (Ministry of Justice, 2018). Strategies with a focus on "prosecutions" (see CPS, 2022) and "bringing offenders to justice" (see Tackling Child Sexual Abuse Strategy, 2021) exist, and whilst such strategies have a clear policy goal, in terms of setting policy agendas and related budgets, there remains a gap in the strategised provision of reintegration support, and allied services. At present, in the community, beyond mandated support from probation services, some of which might include informal economies of care, through their work aligned with probation's role to empower "change" (HM Prison & Probation Service, 2021). However, such a development would also require a simultaneous campaign to advance public education about (sexual) crime and 'justice'. In recent times, there has been a parallel rise in societal discussion, and public campaigns, about global challenges such as climate change (Monroe et al., 2019), however there remains a significant gap in evidence-informed public knowledge relating to (sexual) abuse, which is similarly a "global challenge", and argued to have pandemic-like qualities (Beier et al., 2021). To redress the imbalance there needs to be an effective public health campaign regarding sexual offending, as Laws (2000) put forward fifteen years ago, which would seek to frame private troubles as a public issue (Carrabine et al., 2000). Similarly, there needs to be a parallel campaign and strategy which places emphasis on the importance of principles of "retraction, recency, relevance, redeemability" in promoting fair treatment for people who have a criminal record history (Henley, 2019), demonstrating a commitment to both social justice and human rights (Carrabine et al., 2000). Such a focus would in turn provide a counternarrative to the dominant punitive public discourses. This would thus create a climate for positive change at a meso level.

Structurally, on an organisational level, holistic, strengths-based support operate at the boundaries for people who may have a sexual conviction through to people who may cite unhealthy or problematic sexual thoughts, but largely remains under the purview of several charities, including the Lucy Faithfull Foundation (Bailey et al., 2018) and the Safer Living Foundation (Hocken, 2018), with support, financial or otherwise, being precarious in this area. This is despite both charities delivering much-needed services, from Circles of Support and Accountability (CoSA) (Kitson-Boyce, 2018) for people with a sexual conviction, to The Aurora Programme (Clayton et al., 2022; Hocken, 2018) or the Stop it Now helpline (Grant et al., 2019; Horn et al., 2015), for people who are concerned about problematic thoughts. Beyond these formalised interventions, there is also the need for spaces which provide initiatives that build life and soft skills, which are also often a vehicle to meaningful interaction through distraction. The Safer Living Centre (formerly known as 'The Corbett Centre'), an initiative of the Safer Living Foundation, based in the Midlands, plays a significant role in rehabilitating and reintegrating people who have offence-related thoughts through to those with sexual convictions (Safer Living Foundation, 2023). It does this through the following provisions:

- Supporting access to education, employment and housing;
- Providing assistance with employability skills and identification of job opportunities;
- Offering advice and assistance with supervised internet access and phone calls, for the purpose of communication with potential employers or agencies;
- Hosting lectures and practical workshops on business and financial management skills for those for whom being self-employed post-release is a viable option;
- Providing of a safe and social place for individuals to develop skills such as cooking, mindfulness, and improving social skills;
- Offering a space for religious services and support groups; and
- Support for friends and family is also provided.

Such organisations and initiatives are needed as an antidote to the punitive climate surrounding people associated with sexual crime. This climate is significant upon considering the demands of reintegrating back into the community, of serving a custodial sentence, and having to navigate new realities such a disclosing a criminal record. Thompson and Thomas (2017) discuss in their work the role that CoSA volunteers can have in supporting the 'core' member navigate the disclosure process—"Ashley was applying to begin a course at a local college and was nervous about disclosing his conviction. His volunteers went with him as support" (2017, p. 80), echoing Bates et al.'s (2007) earlier work which found that CoSA helped with a "move to independent accommodation and in the process of disclosure of offences to prospective employers" (Clarke et al., 2017, p. 468). As such, what might be perceived as operating at the margins *matters*, and there is a need to acknowledge the situated nature of disclosure and scaffold the support that exists for individuals who are navigating the perpetual need to disclose their past. As such, there is a need for public discourse to shift, in order for (i) charities to have the support needed to operate on the level/scale required; and (ii) the process of disclosure to not be likened to an "invisible punishment" (Thomas & Hebenton, 2012, p. 238) due to the climate often surrounding it. Support, financial or otherwise, is needed for such organisations which are evidence-informed, committed to social justice and human rights, and seek to connect private troubles with public issues (Carrabine et al., 2000). If this work cannot be done effectively, as discussed earlier, there is the potential for individuals to go unsupported, which can lead to poor mental health and offending. However, a means which may be useful for individuals to seek support to scaffold their wellbeing, and prevent offending, is through social prescribing.

On a micro level, social prescribing (SP) has the potential to embody the principles of public criminology (Carrabine et al., 2000) and seek to redress the gap in connection between individuals with a sexual conviction and the wider community who similarly may have experiences of marginalisation and stigmatisation. This section however is premised on much-needed changes at a macro and meso level to be workable in practice. It is acknowledged that we are in an era where "we are facing a loneliness epidemic" (Holt-Lunstad, 2017, p. 129), and parallel to this, that there is evidence demonstrating the role that SP plays in (i) improving people's health and wellbeing; and (ii) reducing the workload for healthcare professionals and demand for secondary care services (Drinkwater et al., 2019), and thus in England, SP is part of the NHS's Long Term Plan (2019). A heterogenous population who are currently benefitting from social prescribing are those who are older, or elderly. Indeed, it has been reported that "social prescribing has potential to benefit older people by helping those with social, emotional or practical needs to access relevant services and resources within the local community" (Hamilton-West et al., 2020, p. 319). However, it would be erroneous to take a narrow conception of who this pathway would suit. Men with a sexual conviction have the potential to benefit from SP given (i) the established link between having a conviction for a sexual crime and experiencing social anxiety (Nunes et al., 2012; Marshall et al., 2009; Porter et al., 2015); and (ii) the "accelerated *social* ageing process" (Teague, 2022, p. 15) that can occur as a result of risk-focused practitioners, such as probation officers, being able to use their penal power to limit who one associates with and where association occurs, thus unintentionally socially ageing the individual as a result of possibilistic risk, rather than probabilistic risk, being central to consideration (Furedi, 2009). As such, risk underpins sociality and what happens on a micro level, resulting from what happens on both a meso and macro level.

Therefore, in considering the context and climate in which sexual crime is situated in, it is imperative to understand the lived realities of people associated with them, through, for example, having a sexual conviction. Disclosing a criminal record history is not done in an amoral, neutral space. As discussed earlier, there are both individual and societal harms associated with aspects of the disclosure process and the climate surrounding it. In considering (i) public criminology's commitment to connecting public issues and private troubles (Carrabine et al., 2000); and (ii) eleven million people in the United Kingdom having a criminal record (Home Office, 2017), there is a need for the whole of society to problematise the issues around disclosure, and the climate surrounding such. Similarly, there is of course a danger that, in contemplating who, and how many, of the eleven million have a *sexual* conviction, this can fuel a societal binarization of who is "deserving" of support, and who is not (Christie, 1986), thus imposing an unhelpful and harmful hierarchy on people. Such a binary is apparent in the registration requirements of MeetUp (2023a) and further exacerbates vulnerability through parallel processes of stimatisation and ostractisation.

Conclusions

Experiences of (re)integrating into mainstream society are hampered by the social and political climate of the time. At present the dominance of the community/public protection model (Kemshall & Maguire, 2001), and the influence of possibilistic risk (Furedi, 2009), inform the statutory sector and societal response to men associated with sexual crime. As such, this often punitive landscape shapes what sociality looks like, and in what context (Teague, 2022). Given the complex landscape regarding who harms, and who has been harmed, and the "crisis of loneliness" (Holt-Lunstad, 2017) backdrop among other recent crises (e.g., COVID-19), there is a clear need for opportunities to form social connection and community. As such, understanding the opportunities and challenges surrounding these connections is necessary in thinking comprehensively about reintegration. As this chapter discusses, the realities of resettlement are such that processes such as disclosing a criminal record is one which can cause significant distress due to its perpetual nature, and the lack of social provision to counter encountered vindictiveness. Therefore, whilst beyond the scope of this chapter to discuss these in great detail, it is clear there needs to be a review of (i) the policies and processes surrounding disclosure which take into account Henley's work on the importance of the principles of "retraction, recency, relevance, redeemability" for treating people who have a criminal record history fairly (Henley, 2019); and there is also a need to (ii) scaffold social and strengths-based support for people who have a sexual conviction. As discussed previously, there are charities such as the Safer Living Foundation whose approach is strengths-based, and work to support people's holistic needs. The Safer Living Centre (Safer Living Foundation, 2023), for example, is evidence of one such novel approach which seeks to support the development of a "moral community" (Ievins, 2022, 2023) to mitigate the pains and "stain" associated with having a sexual conviction, and the associated challenges. Therefore, it is imperative that policies and practices in the realm of reintegration are informed by the principles of public criminology. Beyond this, charities like the Safer Living Foundation, whilst operating in a tide of populist punitiveness, should operate on more of a statutory basis to ensure a sustained space for support to be provided into the longer term. In the absence of a social justice approach to people with a sexual conviction, there must be stable provision for people who have harmed, and been harmed, to be supported.

References

Alden, L. E., & Taylor, C. T. (2004). Interpersonal processes in social phobia. *Clinical Psychology Review, 24*, 857–882.

Allen, D. S. (2000). *The world of prometheus: The politics of punishing in democratic Athens*. Princeton University Press.

Armitage, R., Wager, N., Wibberley, D., Hudspith, L. F., & Gall, V. (2023). We're not allowed to have experienced trauma. We're not allowed to go through the grieving process—Exploring the indirect harms associated with Child Sexual Abuse Material (CSAM) Offending and its impacts on non-offending family members, *Victims & Offenders*, https://doi.org/10.1080/15564886.2023.2172504

Bailey, A., Squire, T., & Thornhill, L. (2018). The Lucy Faithfull Foundation: Twenty-five years of child protection and preventing child sexual abuse. In R. Lievesley, K. Hocken, H. Elliott, B. Winder, N. Blagden, & P. Banyard (Eds.), *Sexual crime and prevention* (pp. 57–82). Palgrave Macmillan.

Bates, A., Saunders, R., & Wilson, C. (2007). Doing something about it: A follow-up study of sex offenders participating in Thames Valley Circles of Support and Accountability. *British Journal of Community Justice, 5*, 19–42.

Becker, H. S. (1995). Moral entrepreneurs: The creation and enforcement of deviant categories. *Deviance: A Symbolic Interactionist Approach*, 169–178.

Beier, K. M., von Heyden, M., & Schilg, I. (2021). Child sexual abuse as a global challenge. *InPsych*, 43 (3). https://psychology.org.au/for-members/publications/inpsych/2021/august-special-issue-3/child-sexual-abuse-as-a-global-challenge

Bell, K., Winder, B., & Blagden, N. (2018). 'Better as a Buddhist': An interpretative phenomenological analysis of the reflections on the religious beliefs of buddhist men serving a prison sentence for a sexual offence. *Religions, 9*(4), 101. MDPI AG. Retrieved from https://doi.org/10.3390/rel9040101

Blagden, N., Winder, B., Hocken, K., Lievesley, R., Banyard, P., & Elliott, H. (Eds.). (2019). *Sexual crime and the experience of imprisonment*. Springer.

Blagden, N., Winder, B., & Lievesley, R. (2020). The resurrection after the old has gone and the new has come': Understanding narratives of forgiveness, redemption and resurrection in Christian individuals serving time in custody for a sexual offence. *Psychology, Crime & Law, 26*(1), 34–52. https://doi.org/10.1080/1068316X.2019.1634194

Boukli, A. (2017). *Defining Zemia. Harm Evidence & Research Collaborative*. https://oucriminology.wordpress.com/2017/05/08/defining-zemia/

Boukli, A., & Kotzé, J. (2018). *Zemiology: Reconnecting crime and social harm*. Palgrave Macmillan.

Cannon, W. B. (1929). *Bodily changes in pain, hunger, fear, and rage*. Appleton.

Carrabine, E., Lee, M., & South, N. (2000). Social wrongs and human rights in late modern Britain: Social exclusion, crime control, and prospects for a public criminology. *Social Justice, 27*(2), 193–211.

Chatterjee, H. J., Camic, P. M., Lockyer, B., & Thomson, L. J. M. (2018). Non-clinical community interventions: A systematised review of social prescribing schemes. *Arts & Health, 10*(2), 97–123. https://doi.org/10.1080/17533015.2017.1334002

Christie, N. (1986). The ideal victim. In E. A. Fattah (Ed.), *From crime policy to victim policy* (pp. 17–30). Palgrave Macmillan.

Clarke, M., Brown, S., & Völlm, B. (2017). Circles of support and accountability for sex offenders: A systematic review of outcomes. *Sexual Abuse, 29*(5), 446–478.

Clayton, J., Hocken, K., & Blagden, N. (2022). A compassionate intervention for individuals with problematic sexual interests: Group and individual outcomes in the UK. *Abuse: An International Impact Journal, 3*(2), 4–23.

Corda, A., & Lageson, S. E. (2020). Disordered punishment: Workaround technologies of criminal records disclosure and the rise of a new penal entrepreneurialism. *The British Journal of Criminology, 60*(2), 245–264.

CPS. (2022). *Rape strategy update*. February 2022. https://www.cps.gov.uk/sites/default/files/documents/publications/Rape-Strategy-Update-2022-annotated.pdf

Cubellis, M. A., Evans, D. N., & Fera, A. G. (2019). Sex offender stigma: An exploration of vigilantism against sex offenders. *Deviant Behaviour, 2017*(1), 1–16.

Drinkwater, C., Wildman, J., & Moffatt, S. (2019). Social prescribing. *BMJ, 2019*(364), Article l1285. https://doi.org/10.1136/bmj.l1285

Duncan, K., Wakeham, A., Winder, B., Blagden, N., & Armitage, R. (2022). "Grieving someone who's still alive, that's hard": The experiences of non-offending partners of individuals who have sexually offended—An IPA study. *Journal of Sexual Aggression, 28*(3), 281–295. https://doi.org/10.1080/13552600.2021.2024611

Farmer, M., McAlinden, A.-M., & Maruna, S. (2015). Understanding desistance from sexual offending: A thematic review of research findings. *Probation Journal, 62*, 320–335. https://doi.org/10.1177/0264550515600545

Figley, C. R. (1995). Compassion fatigue: Toward a new understanding of the costs of caring. In B. H. Stamm (Ed.), *Secondary traumatic stress: Self-care issues for clinicians, researchers, and educators* (pp. 3–28). The Sidran Press.

Fox, K. J. (2017). Contextualizing the policy and pragmatics of reintegrating sex offenders. *Sexual Abuse, 29*(1), 28–50.

Furedi, F. (2009). Precautionary culture and the rise of possibilistic risk assessment. *Erasmus l. Rev., 2*, 197.

Giddens, A. (1997). *Sociologia [Sociology]*. Fundagio Calouste Gulbenkian.

Goffman, E. (1961). *Asylums: Essays on the social situation of mental patients and other inmates*. Doubleday Anchor Books.

Goffman, E. (1963). *Stigma: Notes on the management of spoiled identity*. Prentice-Hall.

Graber, J., & Zitek, E. M. (2022). Reducing criminal record discrimination through banning the box: The importance of timing and explanation in the reveal of a drug conviction. *Psychology, Crime & Law, 28*(4), 390–415.

Grant, B. J., Shields, R. T., Tabachnick, J., & Coleman, J. (2019). "I didn't know where to go": An examination of Stop It Now!'s sexual abuse prevention helpline. *Journal of Interpersonal Violence, 34*(20), 4225–4253.

Hamilton-West, K., Milne, A., & Hotham, S. (2020). New horizons in supporting older people's health and wellbeing: Is social prescribing a way forward? *Age and Ageing, 49*(3), 319–326.

Henley, A. (2019). Alternative approaches to criminal records: How can we achieve justice as fairness? In P. Carlen & L. A. Franca (Eds.), *Justice alternatives* (pp. 321–339). Routledge.

Henley, A. (2022). Criminalisation, criminal records and rehabilitation: From supervision to citizenship? *Probation Journal, 69*(3), 273–277.

HM Government. (2021). *Tackling child sexual abuse strategy*. https://assets.publishing.service.gov.uk/government/uploads/system/uploads/attachment_data/file/973236/Tackling_Child_Sexual_Abuse_Strategy_2021.pdf

HM Prison and Probation Service. (2021). *The target operating model for probation services in England and Wales: Probation reform programme*. https://assets.publishing.service.gov.uk/government/uploads/system/uploads/attachment_data/file/1061048/MOJ7350_HMPPS_Probation_Reform_Programme_TOM_Accessible_English.pdf

Hocken, K. (2018). Safer living foundation: The aurora project. In R. Lievesley, K. Hocken, H. Elliott, B. Winder, N. Blagden, & P. Banyard (Eds.), *Sexual crime and prevention* (pp. 83–109). Palgrave Macmillan.

Holt-Lunstad, J. (2017). The potential public health relevance of social isolation and loneliness: Prevalence epidemiology, and risk factors. *Public Policy & Aging Report, 27*(4), 127–130.

Home Office. (2017). *Freedom of Information request (our ref: 44921): internal review* [online] 27 October. https://www.whatdotheyknow.com/request/423289/response/1060100/attach/3/44921%20Stacey%20Internal%20Review.pdf

Horn, J. V., Eisenberg, M., Nicholls, C. M., Mulder, J., Webster, S., Paskell, C., Brown, A., Stam, J., Kerr, J., & Jago, N. (2015). Stop It Now! A pilot study into the limits and benefits of a free helpline preventing child sexual abuse. *Journal of Child Sexual Abuse, 24*(8), 853–872. https://doi.org/10.1080/10538712.2015.1088914

Hulley, J. L. (2016). *My history is not my destiny: Exploring desistance in adult male child sex offenders* (Doctoral thesis, University of Sheffield).

Ievins, A. (2020). 'Perfectly individualized and constantly visible'? Lateral tightness in a prison holding men convicted of sex offences. *Incarceration, 1*(1), 2632666320936433.

Ievins, A. (2022). The society of 'sex offenders.' In B. Crewe, A. Goldsmith, & M. Halsey (Eds.), *Power and pain in the modern prison: Revisiting the society of captives* (pp. 175–192). Oxford University Press.

Ievins, A. (2023). *The stains of imprisonment: Moral communication and men convicted of sex offences*. University of California Press.

Ievins, A., & Crewe, B. (2015). 'Nobody's better than you, nobody's worse than you': Moral community among prisoners convicted of sexual offences. *Punishment & Society, 17*(4), 482–501.

Kitson-Boyce, R. (2018). Do circles of support and accountability work? A review of the literature. In H. Elliott, K. Hocken, R. Lievesley, N. Blagden, B. Winder, & P. Banyard (Eds.), *Sexual crime and circles of support and accountability* (pp. 43–69).

Kemshall, H., & Maguire, M. (2001). Public protection, partnership and risk penality: The multi-agency risk management of sexual and violent offenders. *Punishment & Society, 3*(2), 237–264.

Lageson, S. E. (2022). Criminal record stigma and surveillance in the digital age. *Annual Review of Criminology, 5*, 67–90.

Laws, D. R. (2000). Sexual offending as a public health problem: A North American perspective. *Journal of Sexual Aggression, 5*(1), 30–44.

MacNair, R. (2002). *Perpetration-induced traumatic stress: The psychological consequences of killing*. Greenwood Publishing Group.

Mahoney, I., Teague, K., Long, M., & Winder, B. (2022). Populist and vindictive responses to sexual offending against children, pluralities of violence and the implications for criminal and social justice. *Archives of Criminology, 44*(1), 123–145.

Marshall, W. L., Marshall, L. E., Serran, G. A., & O'Brien, M. D. (2009). Self-esteem, shame, cognitive distortions and empathy in sexual offenders: Their integration and treatment implications. *Psychology, Crime & Law, 15*, 217–234.

McAdams, D. P. (2006). The role of narrative in personality psychology today. *Narrative Inquiry, 16*(1), 11–18. https://doi.org/10.1075/ni.16.1.04mcaMcAd

McCartan, K. (2022). *Refining processes in policy and practice in working with people accused or convicted of a sexual offence*. HM Inspectorate of Probation. Academic Insights 2022/01. https://www.justiceinspectorates.gov.uk/hmiprobation/wp-content/uploads/sites/5/2022/01/Academic-Insights-McCartan.pdf

McCartan, K., & Kemshall, H. (2023). Incorporating quaternary prevention: Understanding the full scope of public health practices in sexual abuse prevention. *International Journal of Offender Therapy and Comparative Criminology, 67*(2–3), 224–246.

McCartan, K. F., & Richards, K. (2021). The integration of people convicted of a sexual offence into the community and their (risk) management. *Current Psychiatry Reports, 23*(52). https://doi.org/10.1007/s11920-021-01258-4

McKendy, L., & Ricciardelli, R. (2021). The pains of release: Federally-sentenced women's experiences on parole. *European Journal of Probation, 13*(1), 1–20.
McNeill, F. (2019). *Pervasive punishment*. Emerald Publishing Limited.
McTier, T. S., Jr. (2022). Between a rock and a hard place: Navigating institutional barriers as a graduate student with a criminal record. *American Behavioral Scientist, 66*(10), 1418–1437.
Meetup. (2023a). *Meetup member restrictions*. https://help.meetup.com/hc/en-us/articles/360002913271-Meetup-member-restrictions
Meetup. (2023b). *The people platform: Where interests become friendships*. https://www.meetup.com/
Ministry of Justice. (2018). *A whole system approach for female offenders: Emerging evidence*. Ministry of Justice.
Monroe, M. C., Plate, R. R., Oxarart, A., Bowers, A., & Chaves, W. A. (2019). Identifying effective climate change education strategies: A systematic review of the research. *Environmental Education Research, 25*(6), 791–812.
Morse, D. F., Sandhu, S., Mulligan, K., Tierney, S., Polley, M., Giurca, B. C., & Husk, K. (2022). Global developments in social prescribing. *BMJ Global Health, 7*(5), Article e008524.
Nacro. (2023). *Your criminal record and disclosure* [online]. https://www.nacro.org.uk/nacro-services/advice/advice-for-individuals/understanding-whats-on-your-criminal-record/
Naylor, B. (2005). Do not pass go: The impact of criminal record checks on employment in Australia. *Alternative Law Journal, 30*(4), 174–179.
Naylor, B., Paterson, M., & Pittard, M. (2008). In the shadow of a criminal record: Proposing a just model of criminal record employment checks. *Melbourne University Law Review, 32*(1), 171–198.
NHS. (2019). *The NHS Long Term Plan*. https://www.england.nhs.uk/wp-content/uploads/2022/07/nhs-long-term-plan-version-1.2.pdf
Nunes, K. L., McPhail, I. V., & Babchishin, K. M. (2012). Social anxiety and sexual offending against children: A cumulative meta-analysis. *Journal of Sexual Aggression, 18*(3), 284–293. https://doi.org/10.1080/13552600.2010.549243
Office for National Statistics. (2023). *Sexual offences in England and Wales overview: year ending March 2022*. 23 March. https://www.ons.gov.uk/peoplepopulationandcommunity/crimeandjustice/bulletins/sexualoffencesinenglandandwalesoverview/march2022
Piché, J. (2015). Assessing the boundaries of public criminology: On what does (not) count. *Social Justice, 42*(2), 70–90.

Pickett, J. T., Mancini, C., Mears, D. P., & Gertz, M. (2015). Public (mis)understanding of crime policy: The effects of criminal justice experience and media reliance. *Criminal Justice Policy Review, 26*(5), 500–522.

Pijoan, E. L. (2014). Criminal record disclosure and the right to privacy. *Criminal Law Review, 10*, 723–737.

Porter, S., Newman, E., Tansey, L., & Quayle, E. (2015). Sex offending and social anxiety: A systematic review. *Aggression and Violent Behavior, 24*, 42–60.

Pyrooz, D. C., Moule, R. K., Jr., & Decker, S. H. (2014). The contribution of gang membership to the victim-offender overlap. *Journal of Research in Crime and Delinquency, 51*(3), 315–348.

Rennie, A., & Crewe, B. (2023). 'Tightness', autonomy and release: The anticipated pains of release and life licencing. *The British Journal of Criminology, 63*(1), 184–200.

Safer Living Foundation. (2023). *The Corbett Centre* [online]. https://www.saferlivingfoundation.org/what-we-do/adult-projects/the-corbett-centre/

Saunders, L. (2022). The transition from prison to the community of people convicted of sexual offences: Policy and practice recommendations. *Prison Service Journal, 251*, 11–18.

Scottish Development Centre for Mental Health. (2007). *Developing social prescribing and community referrals for mental health in Scotland*. Retrieved from www.gov.scot/Resource/Doc/924/0054752.pdf

Serpa, S., & Ferreira, C. M. (2019). Micro, meso and macro levels of social analysis. *International Journal of Social Science Studies, 7*(3), 120–124.

Stein, M. B., & Stein, D. J. (2008). Social anxiety disorder. *The Lancet, 371*(9618), 1115–1125.

Stewart, R., & Uggen, C. (2020). Criminal records and college admissions: A modified experimental audit. *Criminology, 58*(1), 156–188.

Tankebe, J., & Liebling, A. (2014). Legitimacy and criminal justice: An introduction. In *Legitimacy and criminal justice: An international exploration*. OUP.

Teague, K. (2022). *'If you were on your own it wouldn't be a thing': Accelerated social ageing in men who have a sexual conviction* [online]. Howard League for Penal Reform. Early Career Academics Network Bulletin. August 2022. Issue 51. https://howardleague.org/wp-content/uploads/2022/08/ECAN-summer-2022-FINAL.pdf

Teague, K., Blagden, N., Winder, B., & Hamilton, P. (2021). *'It hasn't had much effect on me... social contact has been limited since arrest': Part 2* [online]. ATSA Blog. 14 May 2021. https://blog.atsa.com/2021/05/it-hasnt-had-much-effect-on-me-social_14.html

Teague, K., & Winder, B. (2023). *How does the "war on woke" perpetuate harm towards those with a criminal conviction?* [online]. The BSC Blog. 5 January. https://thebscblog.wordpress.com/2023/01/05/how-does-the-war-on-woke-perpetuate-harm-towards-those-with-a-criminal-conviction/

Thomas, T & Hebenton, B. (2012). Dilemmas and consequences of prior criminal record: A criminological perspective from England and Wales. *Criminal Justice Studies, 26*(2), 228–242. https://doi.org/10.1080/1478601X.2012.715583

Thomas, T., & Hebenton, B. (2013). Dilemmas and consequences of prior criminal record: A criminological perspective from England and Wales. *Criminal Justice Studies, 26*(2), 228–242.

Thompson, D., & Thomas, T. (2017). *The resettlement of sex offenders after custody: Circles of support and accountability*. Routledge.

Tyler, I. (2020). *Stigma: The machinery of inequality*. Zed Books.

Unlock. (2023). Self-disclosure statements (often referred to as a "disclosure letter") [online]. https://unlock.org.uk/advice/self-disclosure-statements/

Von Hentig, H. (1948). *The criminal and his victim: Studies in the sociobiology of crime*. Schocken Books.

Ward, T., & Stewart, C. A. (2003). The treatment of sex offenders: Risk management and good lives. *Professional Psychology: Research and Practice, 34*(4), 353.

Weaver, E. (2018). *Time for policy redemption: A review of the evidence on the disclosure of criminal records*. The Scottish Centre for Crime and Justice Research.

Western, B., Braga, A. A., Davis, J., & Sirois, C. (2015). Stress and hardship after prison. *American Journal of Sociology, 120*(5), 1512–1547.

5

The Transition from Prison to the Community for People with Sexual Convictions: Do We Want Them to Succeed or Fail?

Lynn Saunders

Introduction

The number of people in prison in England and Wales with a current or previous conviction for a sexual offence has risen significantly in recent years, partly due to growth in the number of convictions for historical offences over this period, but also due to growing recognition and understanding of the harm caused by sexual offending. As a result, there have been numerous changes in legislation, and developments in professional practice. The criminal justice system struggles to cope with the increased demands of surveillance and other restrictions placed on people with sexual convictions, and arguably, the wider community needs to have more responsibility for public safety by supporting and controlling those at risk of reoffending (Armstrong et al., 2008).

L. Saunders (✉)
University of Derby, Derby, UK
e-mail: l.saunders@derby.ac.uk

N. Blagden et al. (eds.), *Sexual Crime and Community Reintegration*, Sexual Crime,
https://doi.org/10.1007/978-3-032-06034-1_5

However, a significant amount of public and media hostility towards people with sexual convictions potentially creates an inhospitable environment for their rehabilitation. Both the media and the public have a heightened awareness of this feared and stigmatised group. Brown et al. (2008) found that there was a "high level of pessimism in relation to the reform and rehabilitation of sex offenders" (p. 266). This view has implications for the development of policy, the work of practitioners, and for employers' attitudes; it therefore also has implications for the prospects of an individual attempting to gain suitable employment upon release from prison.

The majority of these people will return to live in the community, so it is vital that all aspects of their resettlement are supported. This chapter considers the importance of employment and purposeful activity in this resettlement journey. Although many studies have highlighted individuals' experiences of desisting from crime and the factors that helped support them, relatively little research has focused on people with sexual convictions. Factors that are thought to aid desistance include family support, accommodation, and employment, which are more problematic for people convicted of sexual offences. Although people with sexual convictions have very low reoffending rates, research must explore the best methods for reducing reoffending and supporting people to become or return to being positive citizens who are accepted by their communities.

As part of a PhD project at the University of Nottingham,[1] several people convicted of sexual offences were interviewed in prison and again three months after their release, and their expectations and actual experiences of transitioning from prison to the community were explored. The nature and extent of public-protection measures were investigated, as were the ways in which these impacted both the practicalities of finding work and employers' attitudes towards these people. With reference to quotes from the participants, this chapter presents a summary of some of the results of this work and its implications for policy. Specifically, two of the themes that emerged from the data—"Employment and purposeful

[1] For more details of the study, along with its methodology and methods, please see Saunders (2021).

activities" and "Stigma and challenges"—are examined. First, an outline of the legislative and administrative framework is given, and the established literature on desistance, risk assessment and risk management, are examined.

Legal and Administrative Provisions for Managing Sexual Offending

> Despite the differences in their jurisprudence, governments across continents share the same duty of punishing sex offenders appropriately and aim for the same goal of rehabilitation in society. (Zara & Farrington, 2016, p. 255)

Since the 1990s, in both North America and England and Wales, there has been an increase in legislative and administrative measures to control and monitor people convicted of sexual offences. As Kemshall and Britain (2001) observe, "The 1990s witnessed a growing preoccupation around public protection, with policy and subsequent legislation shaped by separate, although often overlapping, imperatives" (p. 4). They noted that significant drivers for this increase in legislation were protecting the public from severe physical and psychological harm being caused by those committing violent offences and the desire to respond more effectively to an increase in child sexual abuse.

Measures restricting and controlling people with sexual convictions were included in the Criminal Justice Act 1991; these measures allowed for the imposition of a custodial sentence to prevent "serious harm" to the public. The legislative provisions in England and Wales were extended by the Sex Offenders Act, 1997. Part 1 of this act established the arrangements for notification requirements, informally known as "the sex offender register", designed to monitor and track people with sexual convictions. Multiple Acts of Parliament were added over subsequent years, increasing the number and type of restrictions placed on people with convictions for sexual offences. Multi-Agency Public Protection Arrangements (MAPPA), defined as "the process through which police, probation, and prison services work together to manage the risks posed by violent and sexual offenders living in the community in order to protect the public" were extended by the Criminal Justice Act, 2003.

Desistance and Sexual Offending

Desistance research focuses largely on acquisitive and violent crime and the factors that support people to desist from further offending (Maruna, 2001; McNeill, 2006; Weaver, 2015), with the main factors being family support, accommodation, and employment (Brown et al., 2007; Laws & Ward, 2011; Maguire & Nolan, 2012; McAlinden, 2009). Few studies have investigated desistance factors for people convicted of sexual offences and whether their experiences are different from or similar to those convicted of other types of offences. An exception to this is Farmer et al. (2015), who concluded that because of low rates of reoffending, "most people who have committed sexual offences … appear to desist from further sexual offending" (p. 322).

People convicted of sexual offences are often perceived as a homogeneous group (McAlinden, 2006). However, a wide range of offences are considered by statute as sexual offences. These include child abuse, sexual assault, rape, sexual murder, indecent exposure, and internet-related offending. Victim types can vary (children, adult males, adult females), as can the means (coercion, grooming, or surreptitious). Further, offences may be contact or non-contact in nature. It is, therefore, difficult to generalise about the typical "sex offender", as reconviction and reoffending rates—and therefore the risk to the public—are dependent on the range of victim and offence types (National Offender Management Service, 2016).

The differences in the types of offending and the development of an overarching legislative and administrative framework to control and monitor people with convictions for sexual offences—and in some cases to control those who are simply suspected of having committed them—has raised concern regarding the effectiveness of this type of legislation, particularly in the United States. For example, Levenson and Cotter (2005) and Tewksbury and Jennings (2010) have expressed concern about the unintended consequences of these restrictions. Prescott and Rockoff (2011) further argue that these restrictions may adversely impact recidivism, as the people affected become more isolated and detached from communities (p. 4). Tewksbury and Levenson (2007) suggest that restrictions on an individual and the requirement to register and reside in certain areas are fundamentally flawed for three reasons: firstly, media attention on sexual crimes gives a false impression of the scale of sexual crime and that reoffending rates are high; secondly, reoffending rates differ based on offence type but are generally much lower than for people with non-sexual convictions; and thirdly, laws restricting where people convicted of sexual offences may live have been inspired by crimes committed by perpetrators who were strangers to their victims, when in reality, the majority of victims are known to the perpetrator.

Research in North America and the United Kingdom has consistently found that children and young people are much more likely to be victimised by trusted caretakers and relatives than predatory strangers. For example, Radford et al. (2011) found that 90% of children who were abused were victims of someone they knew. Therefore, most of the legislation and its associated restrictions do not significantly reduce the likelihood of the majority of sexual violence or abuse of children and young people. Tewksbury and Levenson (2007) further comment that the legislation's restrictions "lead to instability, transience, and hopelessness, and therefore contradict decades of criminological research identifying factors associated with successful reintegration. Sex offenders with positive support systems are less likely to offend than those that lack support." (p. 56).

Employment and Desistance

A link between work and rehabilitation has been assumed throughout the past century (Gee & Elder, 1986). More recently, the role of education in the reform of individuals has become a key feature of government plans to reform and rehabilitate (Crone, 2016). According to Simon (1993), "Wherever you look in the development of modernist penalty, you will find labor [sic]. Exhort the offenders with religious tracts but make them work... Educate them as citizens but make them work. Treat their pathological features but make them work." (p. 39).

It is often more difficult for people with a criminal history to gain employment when they leave prison (Visher et al., 2011). There are several reasons for these difficulties: individuals may have a poor employment history, substance-misuse problems, or poor educational attainment; in some cases, they may also become deskilled by the very nature of being imprisoned. This is particularly true of people involved in highly skilled occupations prior to their incarceration (Mears et al., 2013). Despite employment being the consistent theme to support desistance from further offending, finding suitable employment can be a significant challenge to people leaving prison with a criminal conviction (Harris & Keller, 2005; Heinrich, 2000). This has significant implications for policy development and demonstrates the need for proactive engagement with employers to improve the employment prospects of people with convictions.

Over time, initiatives in prisons have attempted to provide people with skills to improve their post-release employability. However, Morris (2010) comments that "there is no empirical evidence that teaching someone a trade such as plastering or bricklaying whilst in prison reduces that person's involvement in crime at the end of his or her sentence". Indeed, Morris goes on to say that most policy-linking work with desistance is based on "intuition" (p. 218). Despite this, employment has remained a key component of government policy in reducing reoffending (Ministry of Justice, 2013). This analysis is potentially simplistic for several reasons: some people convicted of sexual offences have offended in the course of their work, or at least whilst in employment; some people do not have the intellectual ability or mental stability to maintain regular employment; and some people are too ill or too old to work and yet still go on to reoffend.

Methods

The findings presented in this chapter are based on interviews with a small group of people in prison at a specialist treatment site for people who have committed sexual offences. Initial interviews were conducted three months before they left prison, and follow-up interviews were conducted up to three months after they returned to the community. The Offender Managers of the participants were also interviewed. At the time of the research, the prison, HMP Whatton, delivered a range of treatment programmes and had the capacity for 841 adult men between the ages of 21 and 89. All had been convicted of at least one sexual offence.

Thematic analysis was used to review the findings of the research. Braun et al. (2014) observe that "thematic analysis can be used to address most types of qualitative research, ranging from questions about individual lived experience through to those about social construction and meaning". Semi-structured interviews were chosen as the preferred method of data collection. Barriball and While (1994) summarise the advantages of semi-structured interviews over other collection methods. They argue that this method is well-suited to exploring attitudes, values, beliefs, and motives, and it gives the opportunity to evaluate a respondent's answers by observing non-verbal cues (p. 329). There is little published research on the experiences of those affected by the transition from prison to the community; therefore, the insights from the people interviewed are particularly illuminating. Ethical approval for this research was obtained from the National Offender Management Service (now HMPPS) and the University of Nottingham.

Participant Selection

At the time of research, HMP Whatton had formed several consultative and focus groups involving prisoners, and these were being used by prison management to consult on potential changes to policy and regime. Researchers were also using a service-user group to discuss existing and potential new research projects. A meeting was arranged with this service-user group to discuss the best way to recruit suitable research participants. Discussion ensued about the merits of using the two main communication tools between prisoners and prison management: the prison had an in-house radio station run and managed by prisoners, and there was also a prison magazine, published quarterly, which was prisoner-led and -managed.

The group proposed that the purpose and objectives of the research should be published on posters in each prison wing, and further information should be provided to prospective participants via a recorded interview on the prison radio station and an article in the prison magazine. The group also suggested that Offender Supervisors (case managers assigned to each prisoner) should be briefed about the study so they could discuss the possibility of participation with eligible prisoners.

A list of prisoners due for discharge over a four-month period was obtained from the prison database. During this period, the prison discharged around 200 prisoners directly into the community. This is a comparatively small number, as the prison housed predominantly longer-term prisoners; therefore, the potential pool of participants was small. However, many prisoners volunteered to be part of the study, and initial interviews were arranged. Because the prison receives individuals with sexual convictions from across the country, follow-up interviews occurred across a wide geographical area. All of the participants who were interviewed in the prison were re-interviewed in the community after their release. This meant that their expectations of life in the community could be compared with their actual experiences. The Offender Managers of the participants were also interviewed, either in person or by telephone.

Experiences of the Participants

The majority of the study's participants were employed at the time of their offending; therefore, the role of employment or constructive purposeful activity after their release potentially has a focus other than simply preventing reoffending: for example, structure, relief from boredom, and improved self-esteem.

One of the most prominent features found during the research was the challenges faced by people in the study leaving prison with sexual convictions in finding suitable employment or purposeful activity. All of the participants in the study had worked throughout their lives, although two had significant health problems during their imprisonment that continued upon release and prevented them from working. However, their licence restrictions significantly impacted their ability to find positive ways to safely spend their time. The structure, self-esteem, and financial benefits of employment or purposeful activity were highlighted in pre- and post-release interviews. However, these were emphasised to a greater extent when the participants were released back to the community.

One participant spoke of the impact that joblessness had had on him both emotionally and financially:

> Extract 1
> I am tearful…The PPU [Public Protection Unit] are still to discuss what types of employment I can do, so I am unable to look for any employment to fund myself.

Another outlined the difficulties he faced finding work specifically because of his convictions for a sexual offence:

> Extract 2
> The National Careers Service gave me a paper, you think, brilliant, I can rely on these people. I managed to get through to somebody and they said we have work in construction. Then they said, 'What's your criminal conviction?' I told them and then they said 'Oh no we can't do anything for you Bye'… I understand it is a criminal conviction, but this could destroy people. They'd be thinking, 'How am I ever going to get a job?'

Several participants spoke of their frustrations in not being able to obtain satisfactory employment, particularly the restrictions placed on them, and the reluctance of Offender Managers and employers to think flexibly about how to safely overcome these restrictions:

> Extract 3

> I think other people find restrictions harder to work with, if they are someone else's they see all of the challenges instead of saying 'ok these are the challenges how can we work with them'.

Another commented more generally on the challenges that other people had in finding employment:

> Extract 4
> I mean I was lucky here because I have got no curfew, I can go out after seven o'clock in the morning and I don't have to be back until 11 o'clock at night, some of 'em [other people living in the hostel] have got four or five curfews in the day so I can understand that it would be a problem getting a job.

Offender Managers who were part of the study also expressed concern about the challenges of people with sexual convictions leaving prison trying to gain employment:

> Extract 5
> There is a stigma attached to the conviction, the stakes will be stacked against them, and even sort of factory jobs will be difficult to secure. The frustration, the aggravation, the upset, the rejection after a time becomes... I can see it wearing on the individuals.

A key finding of the study—that people with sexual convictions have difficulty finding suitable employment or purposeful activity—was underlined by the Offender Managers. All recognised that the requirement to disclose an offence to a prospective employer impacted their ability to find suitable employment, and on the occasions when participants did find work, they often felt anxious about the consequences of their work colleagues discovering their conviction.

Employers' Willingness to Employ People with Convictions for Sexual Offences

The employment of people returning to the community from a prison sentence has been a key tenet of government policy in England and Wales, designed in recent years to reduce crime and criminality (House of Commons Work & Pensions Committee, 2017). However, the Chartered Institute of Personnel and Development (2010) found that 19% of employers said they were likely to exclude people with convictions from the recruitment process. Brown et al. (2007) found that this rose to 50% for people convicted of a sexual offence. Rolfe (2001) suggested that the moral code of the employer often determined their unwillingness to employ people with sexual convictions. However, Brown et al. (2007) found that employers were more concerned about the practical issues surrounding the safety of other staff. Research by the charity Working Links (2010) found that many employers erroneously believed they could not employ people with convictions for legal or insurance reasons. This confusion is an understandable concern, as employers have a legal responsibility to ensure the safety of their employees. The Health and Safety at Work Act 1974, Part I, requires employers to secure the "health, safety and welfare of persons at work" and places a duty on them "to ensure, so far as is reasonably practicable, the health, safety and welfare at work of all [their] employees." Employers may be concerned that other employees may be at risk of harm from people with sexual convictions.

However, not all people with sexual convictions pose the same level of risk. Although it may be unwise for some to be placed in certain workplaces, the majority will not pose a risk to work colleagues' health, safety, and welfare, as research into risk factors and likelihood of reoffending shows. There are significant restrictions on the number of jobs available for people with sexual convictions due to legislation, MAPPA, and licence conditions. This, combined with employer reluctance, significantly limits opportunities for people leaving prison to experience the positive benefits of employment both on their mental wellbeing and financial security. In this regard, the findings in this study support the extant literature. Although six of the participants had skills, a good employment history, and were strongly motivated, employers were reluctant to consider them for jobs within their organisations.

Work and Wellbeing

> Work involves the application of physical or mental effort, skills, knowledge, or other personal resources, usually involves commitment over time, and has connotations of effort and a need to labour or exert oneself. (Warr, 1987, quoted by Waddell & Burton, 2006, p. 4)
> Wellbeing is the subjective state of being healthy, happy, contented, comfortable, and satisfied with one's quality of life. It includes physical, material, social, emotional ('happiness'), and development & activity dimensions. (Waddell & Burton, 2006, p. 4)

The benefits of work and purposeful activities for the broader community are widely acknowledged. Waddell and Burton (2006), in their meta-analysis of the benefits of work, concluded that work "meets important psychosocial needs in societies where employment is the norm" and "is central to individual identity, social roles, and social status" (p. vii). The benefits of the role and impact of work in supporting the desistance journeys of people with convictions other than sexual offences are widely acknowledged. However, the participants in the study experienced pronounced difficulties in obtaining suitable—or in some cases, any—employment, despite their motivation and previous employment history, skills and qualifications, as illustrated by the following quote:

> Extract 6
> I have had absolutely nothing yet. I have applied for all sorts of different things; the Job Centre are trying to help me. They know about my conviction. It is obviously difficult to work around that to find something suitable for me.

The importance of work in increasing life satisfaction was recognised by the Ministry of Justice in a report that acknowledged its role in encouraging desistance and that "engagement in meaningful activity increases life satisfaction" (2013, p. 9). The participants in the study recognised the importance of employment on their wellbeing: one realised the impact that a long period of imprisonment had had on his work skills after his release from prison. He had been promised a job by his previous employer on release but found that after ten years in prison, he was not prepared for the world of work. Another spoke of the difficulties in obtaining and keeping employment because of the restrictions of MAPPA and licence conditions:

> Extract 7
> You can only have jobs that the probation will let you have. I can see why they are doing it but sometimes they go overboard. They seem to be stopping people getting jobs.

From their studies in Canada, Helliwell and Huang (2010) highlight the importance of nonfinancial aspects of the workplace, which they define as "workplace trust". Robinson (1996) defined trust as a person's "expectations, assumptions, or beliefs about the likelihood that another's future actions will be beneficial, favourable, or at least not detrimental to one's interests" (p. 576). One of the participants, for example, highlighted the importance of his role in the prison as a peer supporter to older adults, saying, "it is nice to be trusted". Fine and Holyfield (1996) also comment on the importance of trust, observing that: "There is a world of cultural meanings, emotional responses, and social relations, one not only thinks trust but one feels trust." (p. 25).

Multiple job roles are restricted for people who are subject to notification requirements. These are: working as a teacher or for a children's nursery, in airports or for airlines, in banks or financial institutions, in the health and social care sector, in licenced premises, in roles subject to security-industry licencing, taxi licencing, and joining the armed forces (Nacro, 2018). Brown et al. (2007) argue that "the introduction of the menu of preventative orders introduced to provide a greater level of public protection has, it can be argued, increased the barriers to employment experienced by sex offenders" (p. 33). They also argue that "there has to be a balance between assessing and managing the risk presented by the offender ... There is a political need for sex offenders to be seen to be controlled, and this contributes to the acceptance of increasingly exclusionary and punitive strategies, which result in the barriers to employment becoming more firmly entrenched." (p. 34).

Kramer (1999), however, helpfully summarises the root of the issue surrounding the employment of people with sexual convictions in his study about the nature of trust and distrust in organisations: "Individuals' judgements about others' trustworthiness are anchored, at least in part, on their a priori expectations about others' behaviour." (p. 576). The participants' experiences support this view, as they experienced issues of mistrust not only from prospective employers, but also from Offender Managers and police officers in assessing their suitability for employment. Waddell and Burton (2006), in their meta-analysis of the research into the impact of unemployment, found that there is a "strong positive link" (p. 11) between unemployment and increased rates of overall mortality from cardiovascular disease, lung cancer and suicide, poor physical health, poor general health, poor mental health and psychological wellbeing, and higher levels of medical consultation. They also concluded that the impact of unemployment could be modified by socioeconomic status, gender, family status, age, education, and social capital (p. 12). This is particularly relevant when considering the impact of a period of imprisonment on the participants and the challenges they faced in obtaining suitable employment upon their release.

A large percentage of the wider prison population is unemployed when released from prison: only 27% of people leaving prison said they had a job to go to (Ministry of Justice, 2015). This has wider implications for both the criminal justice system and the economy as a whole. Economists have emphasised the income and consumption consequences of unemployment (Bentolila & Ichino, 2000; Browning & Crossley, 1999), while other research has considered the mental and emotional challenges of unemployment (for example, Argyle, 1999; Darity & Goldsmith, 1996; Frey & Stutzer, 2002). There have, however, been fewer studies examining the impact of unemployment on an individual's wellbeing. Ahn et al. (2004) considered this issue in their study across multiple European countries, and concluded that "unemployment has negative psychological consequences, including loss of identity and loss of self-esteem" (p. 1). Gallie (1999) found that the social isolation that is often a consequence of unemployment is also detrimental to the wellbeing of the unemployed. It is worth reflecting that many participants cited social isolation as the cause of their offending. One participant from the present study articulated his feelings of isolation and loneliness upon his release from prison, so much so that he commented that he would have preferred to be back in prison, fulfilling boring and monotonous tasks but surrounded by others.

The lack of employment opportunities may have more wide-reaching consequences. Ahn et al. (2004) also considered the impact of social isolation on an individual's wellbeing. They concluded that "social relation helps to ease the pain and stress from unemployment" (p. 12). This lack of opportunity to associate with other people because of the restrictions associated with licence conditions and MAPPA, as narrated by two of the participants, together with an absence of opportunity for meaningful employment, is an added burden of a return to the community. Ahn et al. (2004) further considered the differential impact of unemployment on people who enjoy work. They argue that "people choose or do things in order to be happier" (p. 4). They suggest, for example, that people who value money are likelier to do things to become richer, and people who enjoy working are more likely to be employed than those who do not. Therefore, the effect of involuntary unemployment has a greater impact on the wellbeing of this group. Several participants commented on how work was an important part of their lives and how they enjoyed it. Therefore, the licence and MAPPA restrictions in place had a differential impact on their wellbeing compared to others who do not place the same value on work or enjoy it as much.

Clark and Oswald (2002) considered the substantial adverse effects of unemployment on wellbeing in their longitudinal study of the wider population. The specific long-term impacts of unemployment on people with sexual convictions have yet to be considered. However, given the short-term effects apparent in this study, this topic merits further scrutiny.

The challenges of finding acceptable employment for people with sexual convictions need to be addressed, as employment can assist in helping to improve a person's wellbeing and, therefore, is likely to assist in reducing the risk of repeated offending. This study has indicated that efforts need to be made to ensure that employers are made aware of both the opportunities and risks of employing people with sexual convictions, and they should consider people on an individual basis for employment opportunities.

Atonement

> Atonement is both a goal and a process. As a goal, atonement seeks the reconciliation of the wrongdoer and the victim, and the reintegration of the wrongdoer back into good standing as a member of the community. (Garvey, 2003, p. 304)
>
> For the perfect removal of guilt, then the wrongdoer must make atonement for his wrong act. (Swinburne, 1989, p. 12)

Although the idea of atonement has, for the most part, a religious and biblical component and is considered to be a central part of theology (as discussed in the present context by Blagden et al., 2020), atonement as a secular concept can be defined as making amends for wrong or injury. Cousineau (2011) argued that atonement is closely associated with forgiveness, reconciliation, sorrow, remorse, repentance, reparation, and guilt. Some criminological scholars (Bazemore, 1999; Irwin, 2010; Mair & Burke, 2013) discuss the concept of redemption and its role in the desistance process. These are interesting concepts in the context of the research participants' views about how the wider community perceived them, and how they saw the voluntary activities they had undertaken in prison as a way of making amends for their conduct.

The analysis of these data indicates the importance of providing opportunities, both inside and outside prison, to aid the atonement process. It is, therefore, a key responsibility of both policymakers and professionals to facilitate these opportunities. Three of the participants were anxious to involve themselves in voluntary activities both inside and outside of the prison, not only as a way of spending time, but also to improve their self-worth. One participant had volunteered with the Autism Accreditation programme in the prison, and another had been a volunteer with the Age UK team, providing activities and support to older people in the prison.

Mellor et al. (2009) explore the value of volunteering in the wider community and consider two specific elements: "personal and neighbourhood wellbeing" (p. 146). The concept of personal wellbeing is defined in their study as "people's own views of their wellbeing and involves both cognitive and affective components" (p. 146). They found that the volunteers had higher personal wellbeing than those who did not volunteer in their community. They also found that volunteers were more extroverted, less neurotic, more optimistic and had greater perceived control than non-volunteers. They do, however, point out that this relationship is a complex one, because "individuals with higher levels of wellbeing are more likely to engage in volunteer work than those with lower levels of wellbeing" (p. 145).

Other participants in the research had expressed the desire to help others but had not felt able to do so for various reasons. Two of the participants regretted that they had not trained as "Listeners" (a group of Samaritans-trained peer supporters helping people who are at risk of suicide or self-harm in prison), as they felt the need to offer support to others and to help them through the process of imprisonment. Another had given information to the police about how to access websites hosting images of child sexual exploitation—not only to help him to avoid reoffending (because he could more easily be detected) but also to help the police to detect others doing so. The desire of those in the prison where the research took place to engage in voluntary activities, and to raise funds for charity through sponsored events, is also an interesting phenomenon. It may result from a desire to somehow re-establish their role as a member of the wider community and, therefore, to atone in some way for their wrongdoing.

Licence Restrictions and Notification Requirements

The notification requirements for people convicted of sexual offences were introduced in 1997, and the Violent and Sex Offender Register (ViSOR) database was established between 2003 and 2008. Anyone convicted of a sexual offence must tell the police several personal details including their name (and aliases), date of birth, address, and whether they are living with a child under 18. They must also inform the police within three days if any of these details change. This includes if the person subject to registration obtains a new debit or credit card. Even if none of these details change, the person subject to notification requirements must report to the police once every 12 months. The police public protection team will also periodically visit them at home to check that they are living there; they can visit at any time, and they may also decide to tell others about their convictions.

The decision to provide information to others about an individual's convictions will result from their risk of harm to the public generally, or to specific individuals. For example, if a person has a conviction for a sexual offence against a child and a neighbour in the same building has a child, they may be told about the offence. If a person has a conviction for rape against an adult, the police may decide to tell anyone living in the household about the offence if they consider anyone to be at risk. The police may also decide to tell the employer of a person subject to notification requirements about their convictions, if they believe there to be a high risk to others. One participant commented that he did not feel particularly well prepared for what he faced on his return to the community, even though he had attended the meetings in prison to discuss the issue of disclosure:

> Extract 8
> The disclosure meetings and the resettlement work, DWP, and I don't think… I think it would be fair to say that you are still not really prepared for how difficult it is going to be when you get out.

One Offender Manager reflected on the impact of being subject to notification requirements on the people he supervised:

Extract 9
The frustration, the aggravation, the upset, the rejection, time after time, becomes, I can see how it is wearing on the individuals.

This particular insight summarises the thoughts of many of the research participants. The constant reinforcement of registration restrictions, without any clear evidence of the benefits of the process, is a consistent theme. Although there is evidently a role and purpose for the registration and control of people convicted of sexual offences to ensure that people who have been a risk to the public are monitored and their whereabouts are known, the blanket, indiscriminate approach to registration inevitably means that people who are not sexual predators and who are at low risk of reoffending are often subject to a similar level of scrutiny as those whose lifestyle, behaviour, and other risk factors are of more significant concern. This problem dilutes the effectiveness of the registration process and has resource implications for the police service.

Offender Managers consistently commented on the impact of the conviction and subsequent licence conditions on the people they supervised. For example:

Extract 10
Having a sexual conviction imposes restrictions on people that they may or may not be aware of, or have great difficulty overcoming.

Another noted:

Extract 11
For the public, if you hear that someone is a registered sex offender it's almost, almost the worst offence someone can commit, and in their mind it is probably your predatory sex offender that's dragged someone into a bush on a dark night walking home from the pub.

One commented that the impact of the restrictions amounted to: "A very basic level of survival". Another acknowledged the impact that a prison sentence for a sexual offence had had on his client and his wellbeing:

> Extract 12
> They have gone into prison and they have come out, and they are struggling to get a job and that's what gets people down. They don't have a routine, they are on their own because sometimes family don't want to know them, they are struggling to make friends, if they get a partner they have to come in and speak to us and disclose stuff.

Licence restrictions and notification requirements impacted a number of aspects of the research participants' lives. All had far greater challenges in finding employment because of the restrictions placed on them when they returned to the community. One participant expressed his exasperation about his licence conditions:

> Extract 13
> As much as I want to work, I still can't be with people now, I don't know what I can and can't do. So I can't even look, I just find it so frustrating.

The development of IT skills and/or access to the internet is also a significant problem to be overcome for people leaving prison with a sexual conviction; many have licence restrictions that limit their access to a smartphone or the internet. Although it is unpalatable for the general public, it is critical that people with sexual convictions are appropriately prepared for their eventual release from prison to the community. This is mainly because many critical services, such as applications for Universal Credit, job bulletin boards and websites, and job application processes, are almost exclusively online.

The development of prisoners' IT skills prior to leaving prison and the extension of the Universal Credit pilot enabling people to apply for benefits prior to their discharge from prison and access to recruitment sites, would help to ensure that people are better able to deal with the challenges of access to benefits and applying for work. Whilst access to the internet may be permitted in a public place, the reduction in the number of libraries has affected the number of places where even this is possible. In any case, for security reasons, the disclosure of personal data is not recommended in public places. Whilst in prison, advice and support on negotiating these challenges should be provided, together with a realistic appraisal of the consequences of non-compliance.

Conclusions

In this chapter, the challenges of finding suitable employment or voluntary activities are outlined from the personal experiences of the research participants interviewed as part of the study. These can be summarised as: Offender Manager concerns about the risk of contact with the public; employer reticence to employ people with sexual convictions; and community groups concerned about reputational risks or safeguarding issues.

Employment is a key feature of the desistance process and must be tackled. Whilst individuals must be offered assistance with gaining employment—such as CV writing or interview skills—it is also essential to educate, engage with, and support employers to employ people with sexual convictions, and to provide employers with a more informed, balanced assessment of the risks posed by an individual, by looking at people with convictions as individual people rather than a homogeneous group.

Employers should be encouraged to develop their trust in individuals in an informed and supported way. A Ministry of Justice initiative in 2018 focused on promoting several employers (including Pret a Manger, Halfords, and Balfour Beatty) who had prioritised the employment of people with convictions with a view to encouraging more employers to do the same. However, overall, there has been little specific focus on employing people with sexual convictions (Ministry of Justice, 2018).

In times of high levels of unemployment across the wider population, it is perhaps easier to understand the reluctance to employ people with sexual convictions. However, when there are apparent skill shortages across a range of sectors, it is more surprising that employers are unwilling to take on people with sexual convictions, particularly when the reoffending rates of this group are so low. In addition to educating employers, there may be a role for the voluntary sector to take the lead, and for central and local government to offer advice, support, or incentives to take on and safely manage this group of people. In addition, the education of voluntary-sector and community-based organisations to safely open up their services to people convicted of sexual offences upon release from prison should be a priority. If necessary, organisations should be encouraged to develop appropriate risk planning and advised on the management of different restrictions and requirements. These voluntary activities offer the opportunity to support individuals' desire to atone for their wrongdoing and "give something back". It is vital that people are given the opportunity to participate in voluntary and community groups to enable them to achieve this goal.

The experiences of the participants in this study indicate that the responsibility for reducing sexual reoffending, and successfully resettling people convicted of sexual offences in the community—unpopular and unpalatable as this may seem—rests not solely with the individual convicted of a sexual offence, but with the institutions of the state, with practitioners in the criminal justice system, and with employers and with the wider community. If the risks people pose to others are to be safely and appropriately managed, and if people convicted of sexual offences are to be allowed to return to being active and productive citizens, then changes to legal and professional practice and public perceptions are critical. The perception of people convicted of sexual offences as a homogeneous group of highly dangerous, manipulative, and predatory people must be challenged; such labels should be confined to the very small number of people to whom they apply.

The practical suggestions in this chapter can undoubtedly help with this ambition; however, fundamentally, a broader understanding that people convicted of sexual offences are "people like us"—often with the same needs, desires, and personal issues—is a more significant challenge. It is also clear that consideration of appropriate incentives and guidance is needed to support employers to employ people with sexual convictions, both to aid their rehabilitation and to enable them to give back to society.

References

Ahn, N., García, J. R., & Jimeno, J. F. (2004). *Well-being consequences of unemployment in Europe* (Working Paper No. 29). European Network of Economic Policy Research Institutes. https://www.researchgate.net/profile/Juan_Jimeno/publication/239766106_Cross-country_Differences_in_Well-being_Consequences_of_Unemployment_in_Europe/links/09e4150ff91d58402f000000.pdf

Argyle, M. (1999). Causes and correlates of happiness. In D. Kahneman, E. Diener, & N. Schwarz (Eds.), *Well-being: Foundations of hedonic psychology* (pp. 353–373). Russell Sage Foundation.

Armstrong, S., Chistyakova, Y., Mackenzie, S., & Malloch, M. (2008). *Circles of support and accountability: Consideration of the feasibility of pilots in Scotland* (Report No. 01/2008). Universities of Glasgow and Stirling. The Scottish Centre for Crime & Justice Research.

Barriball, K. L., & While, A. (1994). Collecting data using a semi-structured interview: A discussion paper. *Journal of Advanced Nursing, 19*(2), 328–335. https://doi.org/10.1111/j.1365-2648.1994.tb01088.x

Bazemore, G. (1999). *Restorative justice, earned redemption and a communitarian response to crime.* Florida Atlantic University. https://www.researchgate.net/profile/Gordon_Bazemore/publication/242508584_Restorative_Justice_Earned_Redemption_and_A_Communitarian_Response_to_Crime/links/0f317533eeee75caf0000000.pdf

Bentolila, S., & Ichino, A. (2000). *Unemployment and consumption: Are job losses less painful near the mediterranean?* (CEMFI Working Paper No. 0010). Centro de Estudios Monetarios y Financieros. https://ssrn.com/abstract=256134

Blagden, N., Winder, B., & Lievesley, R. (2020). 'The resurrection after the old has gone and the new has come': Understanding narratives of forgiveness, redemption and resurrection in Christian individuals serving time in custody for a sexual offence. *Psychology, Crime & Law, 26*(1), 34–52. https://doi.org/10.1080/1068316X.2019.1634194

Braun, V., Clarke, V., & Terry, G. (2014). Thematic analysis. In P. Rohleder, & A. C. Lyons (Eds.), *Qualitative research in clinical and health psychology.* Palgrave Macmillan.

Brown, K., Spencer, J., & Deakin, J. (2007). The reintegration of sex offenders: Barriers and opportunities for employment. *The Howard Journal of Criminal Justice, 46*(1), 32–42. https://doi.org/10.1111/j.1468-2311.2007.00452.x

Brown, S., Deakin, J., & Spencer, J. (2008). What people think about the management of sex offenders in the community. *The Howard Journal of Criminal Justice, 47*(3), 259–274. https://doi.org/10.1111/j.1468-2311.2008.00519.x

Browning, M., & Crossley, T. (1999). *Shocks, stocks and socks: Consumption smoothing and the replacement of durables during an unemployment spell* (Working Paper No. 376). Research School of Social Sciences, Australian National University. https://openresearch-repository.anu.edu.au/bitstream/1885/40490/2/WP376.pdf

Chartered Institute of Personnel and Development. (2010). *Creating an engaged workforce.* https://www.researchgate.net/publication/253279910_Creating_an_Engaged_Workforce_CIPD_Research_Report

Clark, A. E., & Oswald, A. J. (2002). A simple statistical method for measuring how life events affect happiness. *International Journal of Epidemiology, 31*(6), 1139–1144. https://doi.org/10.1093/ije/31.6.1139

Cousineau, P. (Ed.). (2011). *Beyond forgiveness: Reflections on atonement*. Jossey-Bass.

Criminal Justice Act 1991. https://www.legislation.gov.uk/ukpga/1991/53

Criminal Justice Act 2003. https://www.legislation.gov.uk/ukpga/2003/44

Crone, R. (2016). *Great expectations: The role of education in penal reform.* History & Policy. https://www.historyandpolicy.org/opinion-articles/articles/great-expectations-the-role-of-education-in-penal-reform

Darity, W., & Goldsmith, A. H. (1996). Social psychology, unemployment and macroeconomics. *Journal of Economic Perspectives, 10*(1), 121–140. https://doi.org/10.1257/jep.10.1.121

Farmer, M., McAlinden, A.-M., & Maruna, S. (2015). Understanding desistance from sexual offending a thematic review of research findings. *Probation Journal, 62*(4), 320–335. https://doi.org/10.1177/0264550515600545

Fine, G. A., & Holyfield, L. (1996). Secrecy, trust and dangerous leisure: Generating group cohesion in voluntary organisations. *Social Psychology Quarterly, 59*(1), 22–38. https://doi.org/10.2307/2787117

Frey, B. S., & Stutzer, A. (2002). *Happiness and economics: How the economy and institutions affect human well-being*. Princeton University Press.

Gallie, D. (1999). Unemployment and social exclusion in the European Union. *European Societies, 1*(2), 139–167. https://doi.org/10.1080/14616696.1999.10749930

Garvey, S. P. (2003). Restorative justice, punishment, and atonement. *Utah Law Review, 303*, 303–304. https://scholarship.law.cornell.edu/facpub/279

Gee, E. M., & Elder, G. H. (1986). Life course dynamics: Trajectories and transitions, 1968–1980 by Glen H. Elder Jr. (Review). *Social Indicators Research, 18*(4), 425–428. http://www.jstor.org/stable/27520686

Harris, P. M., & Keller, K. S. (2005). Ex-offenders need not apply: The criminal background check in hiring decisions. *Journal of Contemporary Criminal Justice, 21*(1), 6–30. https://doi.org/10.1177/1043986204271678

Health and Safety at Work Act 1974. https://www.legislation.gov.uk/ukpga/1974/37

Heinrich, S. (2000). *Reducing recidivism through work: Barriers and opportunities for employment of ex-offenders*. Great Cities Institute.

Helliwell, J. F., & Huang, H. (2010). How's the job? Well-being and social capital in the workplace. *Industrial and Labor Relations Review, 63*(2), 205–227. https://doi.org/10.1177/001979391006300202

House of Commons Work and Pensions Committee. (2017). *Support for ex-offenders: Government response to the committee's fifth report*. House of Commons. https://publications.parliament.uk/pa/cm201617/cmselect/cmworpen/1044/1044.pdf

Irwin, J. (2010). *Lifers: Seeking redemption in prison*. Routledge.

Kemshall, H., & Britain, G. (2001). *Risk assessment and management of known sexual and violent offenders: A review of current issues*. Home Office. https://webarchive.nationalarchives.gov.uk/20110218140705/http://rds.homeoffice.gov.uk/rds/prgpdfs/prs140.pdf

Kramer, R. M. (1999). Trust and distrust in organizations: Emerging perspectives, enduring questions. *Annual Review of Psychology, 50*(1), 569–598. https://doi.org/10.1146/annurev.psych.50.1.569

Laws, D. R., & Ward, T. (2011). *Desistance from sex offending: Alternatives to throwing away the keys*. The Guilford Press.

Levenson, J. S., & Cotter, L. P. (2005). The impact of sex offender residence restrictions: 1,000 feet from danger or one step from absurd? *International Journal of Offender Therapy and Comparative Criminology, 49*(2), 168–178. https://doi.org/10.1177/0306624X04271304

Maguire, M., & Nolan, J. (2012). Accommodation and related services for ex-prisoners. In A. Hucklesby & L. Hagley-Dickenson (Eds.), *Prisoner resettlement: Policy and practice*. Routledge.

Mair, G., & Burke, L. (2013). Redemption, rehabilitation and risk management: A history of probation. *Willan*. https://doi.org/10.4324/9780203805961

Maruna, S. (2001). *Making good: How ex-convicts reform and rebuild their lives.* American Psychological Society.

McAlinden, A.-M. (2006). Managing risk: From regulation to the reintegration of sexual offenders. *Criminology & Criminal Justice, 6*(2), 197–218. https://doi.org/10.1177/1748895806062981

McAlinden, A.-M. (2009). *Employment opportunities and community re-integration of sex offenders in Northern Ireland* (NIO Research and Statistical Series: Report No. 20). Northern Ireland Office & Public Protection Arrangements. https://pure.qub.ac.uk/en/publications/employment-opportunities-and-the-community-reintegration-of-sex-o

McNeill, F. (2006). A desistance paradigm for offender management. *Criminology & Criminal Justice, 6*(1), 39–62. https://doi.org/10.1177/1748895806060666

Mears, D. P., Cochran, J. C., & Siennick, S. E. (2013). Life-course perspectives and prisoner reentry. In C. L. Gibson & M. D. Krohn (Eds.), *Handbook of life-course criminology* (pp. 317–333). Springer.

Mellor, D., Hayashi, Y., Stokes, M., Firth, L., Lake, L., Staples, M., Chambers, S., & Cummins, R. (2009). Volunteering and its relationship with personal and neighborhood well-being. *Nonprofit and Voluntary Sector Quarterly, 38*(1), 144–159. https://doi.org/10.1177/0899764008317971

Ministry of Justice. (2013). *Transforming rehabilitation: A summary of evidence on reducing reoffending.* https://www.gov.uk/government/publications/transforming-rehabilitation-a-summary-of-evidence-on-reducing-reoffending

Ministry of Justice. (2015). *National Offender management service annual report 2014/15: Management information addendum.* https://assets.publishing.service.gov.uk/government/uploads/system/uploads/attachment_data/file/449925/mi-addendum.pdf_-_Adobe_Acrobat_Pro.pdf

Ministry of Justice. (2018). *Employing prisoners and ex-offenders.* https://www.gov.uk/government/publications/unlock-opportunity-employer-information-pack-and-case-studies/employing-prisoners-and-ex-offenders

Morris, M. (2010). Prison therapeutic regimes. In S. Wilson & I. Cumming (Eds.), *Psychiatry in prisons* (pp. 210–222). Jessica Kingsley Publishers.

Nacro. (2018). *How different sectors deal with job applications from people with criminal records.* https://www.nacro.org.uk/resettlement-advice-service/support-for-individuals/jobs-and-volunteering/how-different-sectors-deal-with-job-applications-from-people-with-criminal-records/

National Offender Management Service. (2016). *Commissioning strategy. Interventions & services for people convicted of sexual offences* (Unpublished internal document).

Prescott, J. J., & Rockoff, J. E. (2011). Do sex offender registration and notification laws affect criminal behavior? *The Journal of Law and Economics, 54*(1), 161–206. https://doi.org/10.1086/658485

Radford, L., Corral, S., Bradley, C., Fisher, H., Bassett, C., Howat, N., & Collishaw, S. (2011). *Child abuse and neglect in the UK today.* NSPCC.

Robinson, S. L. (1996). Trust and breach of the psychological contract. *Administrative Science Quarterly, 41*(4), 574–599. https://doi.org/10.2307/2393868

Rolfe, H. (2001). *Barriers to employment foroffenders and ex-offenders: Part two - A review of the literature* (Research Report No. 155). Department for Work and Pensions. https://webarchive.nationalarchives.gov.uk/20050816120000/http://www.dwp.gov.uk/asd/asd5/rport155/Main2.pdf

Saunders, L. (2021). *"It felt like falling off a cliff": The transition from prison to the community of people convicted of sexual offences* [Doctoral dissertation, University of Nottingham]. Nottingham eTheses. https://eprints.nottingham.ac.uk/id/eprint/63139

Sex Offenders Act 1997. https://www.legislation.gov.uk/ukpga/1997/51

Simon, J. (1993). *Poor discipline.* University of Chicago Press.

Swinburne, R. (1989). *Responsibility and atonement.* Clarendon Press.

Tewksbury, R., & Jennings, W. G. (2010). Assessing the impact of sex offender registration and community notification on sex-offending trajectories. *Criminal Justice and Behavior, 37*(5), 570–582. https://doi.org/10.1177/0093854810363570

Tewksbury, R., & Levenson, J. S. (2007). When evidence is ignored: Residential restrictions for sex offenders. *Corrections Today, 69*(6), 54–57.

Visher, C. A., Debus-Sherrill, S. A., & Yahner, J. (2011). Employment after prison: A longitudinal study of former prisoners. *Justice Quarterly, 28*(5), 698–718. https://doi.org/10.1080/07418825.2010.535553

Waddell, G., & Burton, A. K. (2006). *Is work good for your health and well-being?* The Stationery Office. https://www.gov.uk/government/publications/is-work-good-for-your-health-and-well-being

Warr, P. B. (1987). *Work, unemployment, and mental health.* Oxford University Press.

Weaver, B. (2015). Offending and desistance: The importance of social relations. *Routledge.* https://doi.org/10.4324/9781315755915

Working Links (2010). *Prejudged: Tagged for life.* Working Links. https://insidetime.org/wp-content/uploads/resources/Employment/Prejudged-Tagged-for-life.pdf

Zara, G., & Farrington, D. P. (2016). Criminal recidivism: Explanation, prediction and prevention. *Willan.* https://doi.org/10.1093/bjc/azv120

6

Life After Release and the Impact of Community MS and SP

Nicholas Blagden

Support

The words in this chapter were written by those who had transitioned from prison to the community. They were all service-users of the Safer Living Foundation and regularly used the Safer Living Centre. The individuals were asked to write in their own words about their experiences of release and their early reintegration back into the community. Both service-users were also asked about the role of community support and what being a part of the Safer Living Centre meant to them. The purpose of this chapter is not to edit their words but rather let them speak for themselves.

N. Blagden (✉)
Safer Living Foundation, Derby, UK
e-mail: n.blagden@derby.ac.uk

N. Blagden et al. (eds.), *Sexual Crime and Community Reintegration*, Sexual Crime,
https://doi.org/10.1007/978-3-032-06034-1_6

Service User 1 (SP)

It would be worth noting that my release from prison was during the COVID-19 Pandemic in 2020 and that my experience may differ from others released during 'normal' times.

Until two weeks before my release, I had no issues surrounding resettlement, owing to my release address being my Dad's house. This provided me with a stable home and a support network. My Mum lived not too far away also. Since being in prison, my release address had been noted as my Dads (and not the address where I was arrested).

Two weeks before release, I was informed by my Offender Supervisor that I would not be returning to my original address. I agreed, as this was where I was arrested and therefore, owing to my offence, it would be unsuitable. I mentioned that my Dad wanted me to 'come home' and was very happy for me to live with him. At this point, my Offender Supervisor was very supportive of this and could not see any problem with this release.

One week before my release however, my Offender Supervisor informed me that my Dad's address had not been approved, and a reason was not given. They asked if there was anywhere else I could go. I mentioned my Mums, she was also happy for me to live with her.

A couple of days later, I was informed that this address was also not approved, and again no reason given. It was at this point that I was told that because I had given no suitable places to live, I would therefore have to be released to a Probation Hostel. I was not happy about this, as I personally saw no reason why my parent's addresses were not considered suitable.

Two days before my release I was told that a Probation Hostel had been found and this would be my release address. My family and I live in X, but the Probation Hostel they had found for me was in XX. This is 50 miles away from [where I live] and my support network, and 80 miles away from my Offender Manager.

Release day came and I lugged my stuff to reception, signed out and was greeted by my parents in the car park. We drove to XX and found the hostel. The hostel staff rang my Offender Manager, who didn't 'need to talk to me' and would phone me on Monday. The joys of Friday release!

I spent the weekend not really knowing what to do or more importantly what I was not to do. I had no phone or computer access. No bank account or money (apart from the £20 my mum had given me). I had been left in a huge city that I had only been to twice in my life. The lack of internet access was an issue. I went to the bank to open an account, to be told to go online and open one. 'I can't do that,' I replied, and the bank worker just looked at me like I was from another planet.

After a lot of back and forth, I resorted to telling them I had just got out of prison, and I don't have access to anything you are requesting. This was my first experience of disclosing, not my offence, but that I had a criminal record. It felt horrible, and that I had been forced to disclose this to be able to get a bank account or be taken seriously about not having any documents or internet access. Luckily, my business bank account with them was still open and they used that as identification to open a personal account.

I spent 4 weeks in XX, with no family contact (due to COVID), and limited access to my Offender Manager. I was moved to a hostel in X and spent 7 months there. My Offender Manager changed to one from X, this took far too long to happen, and I had regular contact with her after making a complaint about the lack of service/support offered.

When I was released, I had what could be both a rational and irrational fear of people finding out about my offence and treating me differently because of it. It was after a few months when this started to not bother me so much. This may have been because there were occasions when I had to disclose to someone and this is something that gets easier to do over time (easier to say, but maybe not hear). I still feel violated that it is something I have to keep bringing up, sometimes for no apparent reason, and going back through everything, having it always there at the back of my mind, only to be brought forward with the dreaded words 'do you have a criminal record?' There are situations where this isn't asked, but you still feel inclined to 'mention' it.

This would then link to barriers being hit head first when trying to do, what should be, the easy things; finding a home, finding work, but also making friends, joining a sports club or even what name to use on social media. This all comes down to the way people perceive those with a sexual conviction and how Probation or the Police want us to be seen.

Everything seems to boil down to 'tell everyone' and 'let the world know you are a sex offender'. I just want to live a happy, private, productive life, without the world thinking they know my life and that my actions from years ago have branded me an 'undesirable'.

I've been punished. I've been to prison. I've completed programmes. I've been rehabilitated. But that's still not enough for everybody. I know, probably the only one that knows, I will not offend again, in any crime for that matter. But the only way to prove this, is to live a long, restricted life, and die. Then everyone will make note 'oh, he said he was not going to offend again. I suppose he was right. Well done.' Not much point now though.

The Centre

The Centre was something I had never heard of, until my Offender Manager recommended that it could be the place for me to move my own personal projects forward. We had talked a lot about my own passion for the prevention of online image offences and my Offender Manager thought the Centre would be a good place to start.

The manager has been very supportive of me and who I am, and always has the time for a chat, even if he is busy or rushed off his feet. Now, having been going for around two years, I am super glad that I was given the opportunity to be involved with every possible aspect of life within the Centre. I don't think I will ever stop going and I passionately support the work they do and have found great friends for life.

When I first stepped foot in the Centre, I was nervous about what to expect. I never really thought it possible for something like this to run with how the public stigma is around people with sexual offences. Yes, it's run in partial secrecy, but with the support it gets from volunteers, etc., proves that things can be done that go against the mould. I now feel confident and that I fit in with the needs of the Centre. The Centre needs me, just as much as I need the Centre. This is why it works well as a support network, by everyone playing their part, to keep the Centre going and providing a safe and happy future for everybody.

Service User 2 (MS)

I never wanted to leave prison. In Whatton prison I was given respect, treated as a human being, had a sense of purpose and loved the work I did there as I felt that it was helping other inmates.

I was very apprehensive when I left prison and never left the hostel I was in for three days, even then it was not without my 'Circle of Support' encouraging me. Without them I would not have lasted long and probably would have been recalled (but not for a sex offence). Thank you to my Circle.

I did make trips into town but for months thought everyone was looking at me, knew what I had done, I felt guilty all the while, walked with my head down, avoided eye contact and kept glancing back to see if was being followed.

Shouts, bangs and running feet were a constant fear, while waiting in a bus queue there was always the fear that someone might strike up a conversation or sit next to me. I seemed always to be on edge, slept poorly, awake at the slightest noise. I did not feel confident or safe for a very very long time, I still don't feel particularly safe.

All convicted sex offenders have problems to a lesser or greater degree, mine maybe partly self-inflicted but my restrictions seem to put me at a disadvantage like don't use public transport, don't make friends without consulting probation first, report all female contact regardless of reason, etc. At least once a week when I go to bed I think, with a bit of luck, I might not wake up, that's sad.

Maybe you are lucky enough to join the Safer Living Centre with XX and all the fantastic volunteers and the other service users, you may begin to feel safe again, feel life's not so bad after all, remember how to smile again and that you can even have fun. Being at the Centre, you feel life's not so bad after all, and you don't need that mask so much. And after each session, you may feel that you can face the world again, can have a life, and may find that you cope with all your restrictions better and anything the world may throw at you. It's helped me, and my life is a little brighter because of it.

Summary

Re-entering the community is challenging for people with sexual convictions due to personal, social, and societal obstacles (Bows & Westmarland, 2018; Mears & Cochran, 2014). Here the service-users speak, in their own words, on the challenges of re-entry, how anxiety provoking and fear inducing it can be, as well as the need for more support for those with sexual convictions. This is epitomised by service-user 2 who states, "I never wanted to leave prison" and how if it was not for community support then he may have been recalled and gone back to prison. This is an initially hopeless and helpless feeling that was attenuated through community support. As has been posed elsewhere in this book (see Chapter 4), do we want people to succeed or fail when they transition back into the community. It is well known that social capital, belonging and hope are important for desistance from sexual crime (see e.g. Farmer, Beech & Ward, 2012). Yet those with sexual convictions are often isolated and alienated and so opportunities to build social capital are likely to be absent, and support must be provided so they can be developed. Albertson (2021) reflects those opportunities must be taken so the person can build social capital which will itself support desistance. However, there is a real lack of support post release for those with sexual convictions and this needs to change if you want to assist men into pro-social offence free lives and ultimately prevent further sexual offending. We would further argue, that when someone transitions from prison to the community it is incumbent on those working with the individual to consider the strengths and 'goods' of that person and to focus on working compassionately with the individual (Winder & Blagden, 2020).

References

Albertson, K. (2021). *Probation Insights: Social capital building supporting the desistance process*. London: HM Inspectorate of Probation.

Bows, H., & Westmarland, N. (2018). Older sex offenders–managing risk in the community from a policing perspective. *Policing and Society, 28*(1), 62–74.

Farmer, M., Beech, A. R., & Ward, T. (2012). Assessing desistance in child molesters: A qualitative analysis. *Journal of Interpersonal Violence, 27*(5), 930–950.

Mears, D. P., & Cochran, J. C. (2014). *Prisoner reentry in the era of mass incarceration.* London: Sage Publications.

Winder, B., & Blagden, N. (2020). Preventing and reducing sexual abuse. In *Making an impact on policing and crime* (pp. 1–33). Routledge.

7

Sledgehammer Cracking the Nut ...The Experience of Licence Conditions and Impact on Community Reintegration

Nicholas Blagden, Bansi Kotak, Samantha Macmillan, Kellsey McCann, and Jade Lamont

Introduction

Most men with a sexual conviction will be released back into the community on licence. The Licence Conditions and Policy Framework (HMPPS, 2021) states that the aims of a licence period are to be preventative, not punitive, ensure public protection, prevent re-offending and

N. Blagden (✉)
University of Derby, Derby, UK
e-mail: n.blagden@derby.ac.uk

B. Kotak
HM Prison and Probation Service, Leicester, UK

S. Macmillan
University of Nottingham, Nottingham, UK

K. McCann
Cambridgeshire & Peterborough NHS Trust, Nottingham, UK

J. Lamont
University of Nottingham, Nottingham, UK

N. Blagden et al. (eds.), *Sexual Crime and Community Reintegration*, Sexual Crime,
https://doi.org/10.1007/978-3-032-06034-1_7

secure the successful reintegration of the offender into the community. Licence conditions are thus used within the UK to protect the public, to prevent re-offending and to secure the successful re-integration of the offenders into the community (National Offender Management Service (NOMS), 2015). For individuals with sexual convictions within the UK, a set of general licence conditions are applied with additional conditions if deemed necessary. However, there has been very little empirical attention paid to how licence conditions are experienced, the impact they have on people's transition from prison to the community and the role they play in the daily lives of men with sexual convictions under supervision.

As of June 2021, there were 87,550 people in prison in the UK (Sturge, 2021). Of these, around 20% have been convicted of a sexual offence (Ministry of Justice (MoJ), 2020). Since 2010, sentence lengths for those with sexual convictions have increased by over 20%, with an average sentence of five years in prison (Williams & Bailey, 2019). The overwhelming majority (>99%) of these individuals are men, and most will be released into the community. There are approximately 54,000 people living in the community who, as a result of the Sex Offences Act 1997, are required to register with the police because they have committed a sexual offence. It is also estimated that there may be over 50,000 people under investigation for a sexual offence at any one time.

There is ongoing debate as to whether the aims of licence conditions are achieved by the licence process and whether an increase in punitive approaches is actually effective. There is a growing body of literature which asserts that certain conditions may actually impede the reintegration process and instead act as a constant reminder of the punishment society believes is apt for certain offender groups (McCarthy & Brunton-Smith, 2018). Owing to the nature of their crimes, individuals who have committed sexual offences typically face more extensive, additional and restrictive conditions than other offending groups, including curfews, monitored relationships, exclusion zones and "no association" rules that limit contact with other offenders. There are concerns that many of the areas where restrictions are most stringent (monitoring of employment, relationships and housing) provide some of the most important protective factors for such individuals (LeBel et al., 2008). Creating barriers that impede individuals' access to 'social capital' has been shown to predict future offending behaviours (Tolson & Klein, 2015), and thus may act in a way that is contradictory to the original ethos of reducing recidivism rates.

Narratives of Re-entry and Sexual Crime

Maruna and LeBel (2003) argue that there are broadly three narratives of community re-entry: control, support (needs-based) and strengths-based. Control narratives are an approach to risk management that favours intense community monitoring, supervision and threat of sanctions (e.g., recall/breaches) as a method of quelling criminality and keeping the public safe. Public safety and protecting the vulnerable from victimisation are inherently good things, and we must protect the public from sexual harm. However, control narratives have largely failed in having any significant impact on recidivism, and there is little evidence that they contribute to any meaningful behaviour change (Maruna & Lebel, 2003). Indeed, determining who is at risk, for how long, and what the response should be are complicated issues, and any response to those with previous convictions needs to be proportionate and procedurally just. There is also a need to understand the context and trajectory of this group. For example, when aggregated across all risk levels, it has been found that the recidivism rate for those with sexual convictions is around 8% (Mews et al., 2017). Low risk offenders meet the desistance threshold at the time of release, and within 10 to 15 years, the vast majority of individuals with a history of sexual crime will be no more likely to commit a sexual crime than individuals who have been convicted of a nonsexual crime and who have no sex offence history. The risk of a further sexual offence declines the longer individuals remain sexual offence-free in the community. It is important then to help people lead offence-free lives. However, there is very little (if any) evidence that imprisonment or intense community supervision deters subsequent criminal behaviour (de Valk et al., 2015). Community notification and associated schemes have been found to have a detrimental and disruptive impact on an individual's reintegration journey and further compound negative psychological consequences (Lasher & McGrath, 2012; Zevitz & Farkas, 2000).

There is a need, then, to go beyond 'control' in order to keep people safe, as constant monitoring is not sufficient. Support or needs-based narratives suggest that those released from prison have multiple deficits and outstanding criminogenic needs (dynamic risk factors) that need intervention in order to reduce crime. These 'needs' are well located within the 'what works' literature and have a sound evidence-base (Mann et al., 2010; Maruna & Lebel, 2003). However, it has been argued that just focusing on 'deficits' or reversing risk factors is not sufficient for lasting change (Ward, 2017). The strength-based narrative aims to redress this and provide a more positive and holistic narrative of re-entry. This narrative asks what people want from life and what contribution do they want to make (Maruna & Lebel, 2003). The strengths-based narrative has parallels with promoting protective factors and the Good Lives Model (GLM) (Ward, 2010). The GLM assumes that all humans fashion their lives around their core values and follow some sort of (often implicit) good life plan. The promotion of protective factors, which Ward (2017, p. 26) defines as "internal and external capacities and personal priorities that enhance iindividuals'well-being and reduce the likelihood that they will harm others or themselves", are thus important for leading offence-free lives. It is therefore important that licence conditions do not unintentionally impede individuals as they seek to transition back into the community, while at the same time ensure that the public are safe.

Barriers to Effective Reintegration

The desistance literature suggests that punitive restrictions are causing barriers to protective factors and hinder the transition from prison into the community (Brown et al., 2007; Tolson & Klein, 2015). One key factor for 'making good' is finding a place in society. Over 99% of people who serve custodial prison sentences for a sexual crime will leave prison and seek to integrate into the community. Waddell and Burton (2016) found that employment was central to people's identities and social status, as work is construed as a way of contributing and belonging to something. In research by Tovey et al. (2022), which focused on people with sexual convictions seeking employment, they found an acute sense of loss from the participants at being denied the opportunity to return to the workforce, and there was a longing for relationships with peers. Social interactions with prosocial peers are a protective factor against recidivism and are of benefit to a person's general psychological well-being and social capital (Harris, 2020). While there are challenges for employers, the stigma of a sexual offence, fears around risk, and what other workers may think/do, there is a general reluctance to hire people with sexual convictions (Tovey et al., 2022). Winder (2022) has delineated 13 factors relevant in reducing sexual recidivism and promoting community integration, two aims that are intertwined. Barriers include: the state of people as they exit prison (in terms of age and health, for example), and the restrictions and constraints upon people in the community. There are also gaps in provision for those transiting from prison to the community, and challenges with reconnecting with family and friends. Some people will also be released into accommodation that is unfit for purpose (see Lomas, 2021). A further barrier is about self-identity—how people perceive themselves, and how others perceive and label them. What we call people has a significant impact on how people think about themselves, and how others think and behave towards them (Willis, 2018). These barriers to successful reintegration often lead to social isolation and prevent desistance from crime being achieved (Tewksbury & Mustaine, 2009).

The aim of this study was to gain an insight into the experiences of people with a sexual conviction as they return to the community, and to specifically explore their experiences of licence conditions, how they navigate such conditions and to understand the impact they have (positive and negative) for those transitioning back into the community.

Method

Participants and Data Collection

Data were collected through the use of a semi-structured interview, each lasting between 45 and 60 minutes. Interviews were conducted in a purpose-built interview room within a support centre or over a secure video link. Participation was voluntary, and all recruited participants ($n = 16$) agreed to engage in the research process. The interview schedule was formed of five separate sections: 'Rapport building', 'Understanding their transition from prison into the community', 'Understanding the impact of licence conditions and living conditions, 'Post-release/reintegration' and 'Hopes for the future. To facilitate discussion, all questions were kept open (Knight et al., 2003). This style of interviewing also enables "rapport to be developed; allows participants to think, speak and be heard; and are well suited to in-depth and personal discussion" (Reid et al., 2005, p. 22).

Data Analytical Procedure

Data were analysed using thematic analysis, a method for identifying, analysing, and reporting patterns and themes within a dataset. Thematic analysis aims to capture rich detail and represent the range and diversity of experience within the data (Braun & Clarke, 2006) and has been described as a 'contextualist method', sitting between the two poles of constructionism and realism. This position thus acknowledges the ways individuals make meaning of their experience, and, in turn, the ways in which the broader social context impinges on those meanings. As such, thematic analyses reflect 'reality' (Braun & Clarke, 2006). The analysis adhered to the phases of qualitative thematic analysis as outlined by Braun and Clarke (2021), consisting of familiarization and detailed readings of the data collected, progressing to initial and systematic coding of the data, and then generating initial themes from the coded data. The final phases included reviewing themes, ensuring that they were consistent with the coding and that they were grounded in the qualitative data (Braun & Clarke, 2021. The final themes were representative of the sample. In this chapter, data extracts are labelled 'quotes', as this term was deemed more humanising than 'extract'. However, extracts/quotes are utilised in the analysis as outlined by Braun and Clarke (2013) and Braun et al. (2016) in which extracts are illustrative of a theme. In this chapter, the quotes used revealed the complex meanings participants had on living with licence conditions (Oliver et al., 2005).

Analysis and Discussion

The superordinate theme and related subordinate themes presented in Table 7.1 are taken from a larger study on the transition from prison to the community for men with sexual convictions. The focus of this chapter and on this aspect of the study is the role and impact of navigating and living with licence conditions as people return to the community.

When performing qualitative analysis, particularly in-depth analysis focusing on participant experiences, the task is to identify the underlying principles that organize the thinking of the participants, and thus the structures that influence perceptions, feelings, and behavior (Skrapec, 2001). The experiences and structures relating to how participants perceived the impact of licence conditions are represented in the subordinate themes in Table 7.1. Each of those themes will be unpacked in the following analysis.

Table 7.1 Superordinate and Subordinate themes

Superordinate theme	Subordinate themes
Experience of licence conditions as a barrier to reintegration	- Setting up to fail - Perceptions of (un)fairness - No Association - Limiting goals and navigating conditions

Subordinate Theme 1 Setting up to Fail

All participants described their transition into the community as often difficult, powerless and entirely controlled by probation. Indeed, the excitement individuals' experience about release, often dissipated into uncertainty and anxiety.

Quote 1
"I was worried of the uncertainty in case something changed at the last minute and it added things. One of my license conditions is that I can't be in sight of any school, swimming baths or children's play area etc. well that's not a homogenous list – it's not obvious what the etc. encompasses. I mean, can I go to a beach? Does that count? If I'm walking… and see a children's play area down the hill like half a mile away can I stand there because that's technically in sight of. Like the high school tram stop for example, if I'm waiting for a tram I'm technically in sight of a school, can I stay there?"

The participant in quote 1 struggles to make sense of the ambiguity of his license conditions, which has amplified his worries of his conditions. His use of rhetorical questions when describing what his daily internal monologue often is, demonstrates his uncertainty and his fear of recall, if he breaks his licence conditions. There is also a sense of confusion due to the lack of clarity determining where he can actually go within the community. This quote suggests he is at a loss in his transition in which can lead him to feeling isolated and frustrated. Understandably, the aim of license conditions is to restrict those with sexual convictions from reoffending in pursuance of protecting public safety (Janicki, 2006). However, all participants mentioned they were aware that license conditions socially and practically limited them, but reflected on the lack of clarity and understanding of their conditions. This in turn, created heightened anxiety and confusion upon their release:

Quote 2
Well my exclusion zone was X, my probation officer was from X and every week I had to go and see her in X so the first day out of prison I was breaking my license conditions under the probation officers orders

because they said to me on the day of release you have to go to X to see your probation officer.

This participant asserts that his experiences and understanding of licence conditions have been unclear and uncertain. This disjunction, in practice, has led him to breaking his license conditions on his first day of release, which understandably, could be distressing when there are consequences of recall for breaking conditions. Furthermore, practice that is not consistent with their license conditions is an additional pressure and stress, which can compound the emotional experience of transitioning back into the community.

Quote3
It's like having a speedo from 0 to 70 and you've no idea where the needle is whether it's on 0 or 70 and there's no indication – you – the landscape is blank… so I think for the purposes of what we're trying to do here I would say the most important thing is this lack of awareness of what comes next and how to deal with it.

Quote 3 further highlights the emotional vacillation experienced by those with sexual convictions as they transition into the community and how a lack of awareness of their licence conditions compounds these feelings. Quote 3 describes their release as being a blank 'landscape', illustrating the lack of control he feels over his reintegration process. There is a loss of focus from what he is trying to achieve upon release and heightened anxiety about the uncertainty of his future in the community. Tewksbury and Copes (2013) found that men with sexual convictions often had little knowledge and understanding of their restrictions upon their release in which leads to feeling a sense of helplessness and uncertainty, which could lead to higher risk of recidivism or poor compliance. For some participants, there was confusion over why certain conditions were on their licence.

Quote 4
No one knows why it's there…I've been asking for two years and I still don't have an answer…the fact you're not told makes it harder. If they

gave me a reason, then yes, I might not like it but I might understand more…so that's another frustration.
Quote 5
Mine were all very well worded and perfectly justifiable. I think that's the other thing, if people look at them and go 'yes that all makes sense and I understand them…and I can see why they're there', then even if you feel upset by them then that's tough.

The extent to which participants could identify with reasons for their licence conditions influenced the extent they perceived them as legitimate. As a consequence, those who felt properly informed; understanding meaning, purpose and implications, appeared to have less "frustration" (quote 4)—as evident from quote 5 who notes his conditions as "perfectly justifiable". This aligns with Ugwudike (2010), claiming such positive experiences bolster belief in authority legitimacy. However, for the participant of quote 4, the lack of fit between the nature of his exclusion condition, his offence and perceptions of risk, coupled with a lack of justification,—influenced a sense of injustice. This only made it "harder" as quote 4 indicates, as grievances persist "2 years" into his reintegration. Interestingly, quotes 4 and 5 both highlight that even if given an undesirable, rationale which they may "not like" or be "upset" by, acceptance would come easier.

Subordinate Theme 2: Perceptions of (Un)fairness

Participants described how they felt that their conditions were not reflective of the crime that they committed, and acted as unnecessary barriers to various factors such as friendships and employment. It was therefore seen that participants felt a desire for their conditions to be personalised, helping them to perceive the licence conditions as fair.

Quote 6
I think a lot of conditions, especially for someone like myself who's on licence for life, I think they bung these conditions on generally. It's not much about the individual – it's not like 'he's done that so we'll add on that' I think they really need to look more at the individual because for

> me I've not acted in a group, the crime was only me ... so yeah I do find it quite restrictive that condition.

Quote 6 highlights the lack of individualisation of the licence conditions for the participant, he felt that this lack of individualisation made his life more 'restrictive' than he felt it needed to be. The participant expressed the need for his licence conditions to relate to the crimes he has committed, and therefore implies the need for his treatment on licence to be fair and appropriate for his crime. This feeling of perceived fairness relates to the procedural justice, as those who believe the licence conditions are fair are more likely to comply with these restrictions and reduce reoffending (Gladfelter et al., 2018). It was an important theme within the data that licence conditions were tailored to the needs of the individual, and it seemed more important that the conditions were seen as fair than the amount of time that he had to abide by them. The participants seemed to emphasise the need to be looked at as an individual, it emphasised the importance that was felt about being viewed as individuals rather than all sex offenders being seen as one. Restrictive licence conditions tend to be 'overinclusive' and exceed the non-punitive purpose of the conditions (Durling, 2006).

> Quote 7
> I just mentioned...'I'm going to stop in a village where my dad's buried and look at the grave' and she went 'no you won't, you will come straight here' and I said 'okay I'll do that' and she added a licence condition of banning me from going to the village...You just don't need to do that. I told you what I was going to do, you said I couldn't and I said I wouldn't – you didn't need to add that.

Quote 7 highlights how participants felt that licence conditions were things that were done to them and that additional conditions were punitive and unfair. Procedurally just treatment is associated with higher outcome satisfaction ratings and decision acceptance, greater cooperation with, and confidence in, the Criminal Justice System, and more law-abiding behaviours (see e.g., Casper et al., 1988). There was a need for the monitoring process to be more collaborative and tailored to individuals, rather than prescriptive or reactive.

Participants shared that whilst they believed they were able to set goals within licence conditions, these goals were adapted to fit within their restricted post-release lifestyle. Some suggested that they felt goals were 'paused' until their time on conditions ended, whereas others suggested boundaries were helpful in guiding realistic goals.

> Quote 8
> But most of the goals I've got, all of my goals, I do set within the perimeter of licence conditions being taken into account, there's no point, you know, to me, there is no point aiming for something knowing full ell the licence conditions will prevent that. Unless, along with that I'm also fighting to get that condition either changed or removed, or whatever *[mhm]* not really, it's limited the goals I think… would be the right wording for me.

Quote 8 describes having to set all goals within the perimeters of his conditions, perceiving them as a physical limiter to achieving his desired goals. Persistent barriers to goal achievement can propagate feelings of hopelessness, with the individual feeling as though they are merely *existing*, this can result in apathy and affect motivation to continue efforts to reintegrate successfully into the community (Nugent & Schinkel, 2016). Throughout the interview data participants expressed a desire for goals to be linked to achieving a prosocial identity, including roles that assist with such identities, e.g., work or volunteer roles, e.g., within the local council and church. Relational distance from the community is a significant pain that comes with desistance, with the most pain being felt by those who wish to engage in goals at the macro-level (Maruna & Farrall, 2004).

Subordinate Theme 3: No Association

Every participant, and seemingly every person on licence for a sexual conviction, will have the additional condition of not to "associate with a known sex offender". The impact of this was discussed by all participants.

> Quote 9

It's mainly the ones [friends] that were sort of from the area and what have ya, the ones that I sort of spoke to quite a bit, and said ah here's me details, if you wanna keep in contact, but then obviously, yeah like as soon as I got out and they put the no association thing on there, it was like well there goes that idea of, 'cause obviously if you're out here then and they're getting ready for release and what have ya, and they're worried about something and you've obviously gone through it yourself then you can't even help 'em.

Quote 9 highlights the experience of not being able to associate with someone whom they had built up a relationship with because they had a sexual offence conviction. The quote portrays a sense of hopelessness and feelings of uselessness as a result of not being able to put personal experiences and wisdom to use. The idea that "you can't even help 'em" highlights how a potentially important resource, i.e., other people with convictions, is not utilised and may make it difficult for some to transition into the community, especially those with a lack of support. This could potentially negatively affect the experience of reintegration because of a lack of social support, and how important this is to desistance and life satisfaction (Maruna & Toch., 2005; Bohmert et al., 2018).

Quote 10
I think the one that causes the most problems is the non-association condition and I know it's caused a massive problem; it often doesn't make sense and I think the classic sledgehammer cracking the nut … I would go to the library and regularly I would see people I recognise from [most recent prison] … now we don't speak we look and go "oh hello", it's in our minds really and actually you know we could say to somebody 'how are you getting on?' … but the non-association clause makes it really difficult.
Quote 11
Sex offenders don't sit and talk about their offences to others; it's just not done because I don't know why it is- bit of self- shame or something… It gets them out and meeting other people and we can talk to each other…Nothing like 'oh you see this girlie down my road' because I wouldn't stand for it, I'd straight away report it.

These quotes highlight how the 'no association' condition is experienced and how, in some circumstances, it can be a relational impediment. Quote 10 highlights how this condition is seen as unnecessarily punitive and something most participants struggled to understand. Quote 11's claim that "sex offenders don't..." has some empirical support, for example, Blagden et al. (2016) and Blagden and Wilson (2020) found that individuals with sexual convictions rarely speak about their offences to others. More specifically, these individuals do not discuss their convictions in order to distance themselves from the 'sex offender' label, which has adaptive and positive consequences. As Blagden et al. (2014) argue, by distancing themselves from the 'sex offender' label, they are seemingly more likely to reject sexual interests in children and identify as a more pro-social individual (Blagden et al., 2014). A further unintended consequence of the no-association condition is relational.

> Quote11
> [the impact] on me, it's not being able to help as much as I want to because I think if, if I could only have just helped that person, he might've not given up, he might've not reoffended, or he, I've, I've gave that person something so that I feel I've helped that person, that's helping me, give me the experience to help.

Linked to other quotes in this theme, the above quote highlights how the "no association" condition impedes the ability of the participants to help others reintegrate and desist from offending. Quote 11 highlights the participant's inability to fill the supportive role for others in similar situations to himself, and it displays feelings of sadness and personal disappointment. Having such roles should not be underestimated and are important in the 'wounded healer' narrative and can be important in the desistance process. Reciprocity is a relational aspect of the desistance process and building on the capacity and opportunity to give back, to gain meaning from shared problems, to reciprocate is an important consideration for criminal justice professionals, especially as it can help individuals enact 'good' selves (see e.g., Perrin & Blagden, 2014). As LeBel et al. (2015) conclude, helping others appears to have adaptive consequences for prisoners and ex-prisoners, and on these grounds, an argument can be made for increasing opportunities for reintegrating ex-offenders to engage in roles characterized by reciprocal helping.

Subordinate Theme 4 Limiting Goals and Navigating Conditions

Participants shared that whilst they believed they were able to set goals within licence conditions, these goals were adapted to fit within their restricted post-release lifestyle. Some suggested that they felt goals were 'paused' until their time on conditions ended, whereas others suggested boundaries were helpful in guiding realistic goals.

> Quote 12
> Umm, yes, yeah I would say it did, um ... whatever it happens to be, if you're looking to do something relatively simple and you're anticipating there's going to be issues sprung up, you kind of question yourself and ask, you know, is it really worth the months' worth of chasing and badgering and that, for this thing or should I just leave it and wait until my licence is done really. So, um, it certainly gave me a pause in terms of pursuing the things I wanted to achieve.

Quote 12 discusses how he anticipated issues arising from simple requests relating to his goals, his language mirrors the frustration expressed in previous quotes. Again, licence conditions are construed as a preventative barrier that inhibits the pursuit of goals instead of supporting individuals. The participant further describes feeling like his life is 'on pause' until his conditions end, limiting his ability to pursue what he really wants to achieve. There is a relational frustration inherent in this quote in the participants' use of "months of chasing", in that some participants took a defeatist (though potentially more pragmatic approach) of not challenging the status quo. The feelings from the participant in quote 12 exemplifies this when he suggests that he just will 'leave it and wait until my licence is done', highlighting the liminal state desisters find themselves in when they are no longer offending but feel unable to fully reintegrate and move on with their non-offending identity (Nugent & Schinkel, 2016; Doekhie et al, 2017). This leads to individuals finding a way to navigate or work with licence conditions.

> Quote 13
> Um…I think there's not a massive impact on them because I tend to choose jobs that I know won't be an issue. But otherwise there's things like, I've previously been retail trained and that sort of thing, my current probation isn't very keen on me working too close to the public, so working in the supermarket is not an issue because lots of people come and go, lots of staff. If I was to sort of work in an independent shop where you know members of the public on a more personal basis, she wouldn't be very keen on that at all. So I basically try to find work and things like that where I'm not working with the public in general.

Quote 13 explores how the participant believes licence conditions have affected his ability to gain his desired employment. Although he discloses that he does not think the conditions have had a 'massive impact', he then describes them as restricting his choice of employment. As a group, these men face significant barriers when trying to regain employment (Rydberg, 2018). Over time, feelings of restricted choice can lead individuals to feel as though they've made few meaningful steps towards their desired identity, thus delaying the process of reintegration (Wagner, 2010).

Discussion

This study sought to understand the process of reintegration for people with sexual convictions, with a particular focus on navigating licence conditions. Developing relationships whilst on licence proved a significant theme. Attempting to establish new relationships whilst experiencing the fear of the 'sex offender' label and the high likelihood of stigma and rejection was a potent challenge. These experiences of *otherness* and ostracisation are consistent with findings within this population (Burrows, 2016). Although this study offers new insight into how the licensee perceives the conditions and how negotiating the label is a constant issue. The data presented here also points to an unintended consequence of licence conditions being a reinforcer of this label, due to it being a primary relational impediment in interactions with others. Some spoke favourably regarding their interactions with professionals, though there was a sense that many relied upon supervisory relationships to fill the void of social isolation from the community. Although empathic staff relationships can foster change within this group (Blagden et al., 2016), organic relationships that offer solidarity, reciprocity, subsidiarity, and a sense of "we-ness" are the most beneficial in fostering desistance in the community (Richards, 2022; Weaver, 2015; Weaver & McNeill, 2015). Such relationships provide deeper interaction than seen in the functional obligations of supervisory support, instead offering super-functional, entirely human interactions that are pivotal for true reintegration into society. Licence conditions appear to perpetuate a state of relational ambivalence, which can only cease when obligations to conditions end. Times of transition can be particularly destabilising for desisters (Digard, 2010), and thus it is concerning that many perceive this to be the only time they would feel able to pursue new relationships.

The participants explored perceptions of (un)fairness with licence conditions, which was experienced as a process done to the participant, rather than with them. There was a breakdown in communication between the participants and the professionals surrounding them, and the feeling that licence conditions were generalised rather than specific. The generalised conditions created underlying feelings that the conditions were 'stupid' and did not aid rehabilitation or reintegration, these experiences suggest a need to create more specific licence conditions for individuals in the system. This is supported by the literature that suggests the increased personalisation or co-production of licence conditions could increase the effectiveness (Weaver, 2015). Research has therefore suggested that a shift towards personalisation of licence conditions would encourage both fair treatment of individuals with convictions, whilst also effectively protecting the public from any risks (Weaver, 2015). The importance of communication and the relationship between the individual and their probation officer was also key for rehabilitation into the community. This relationship should therefore be stressed as important to the probation team, to ensure the individuals feel like they are being fairly treated, as well as a way to build bonds to the community supporting desistance.

Licence conditions are currently delivered as a tool of risk monitoring, these findings provide further evidence that challenge the legitimacy of such an approach. Future research should pay particular attention to what constitutes clear and individualised delivery of licence conditions. Also, exploring how conditions can be used to support the development of protective relationships and goals. There are several limitations to consider within this study. The sample size was relatively small for research of this nature (Guest et al., 2006), and it is important to consider that these participants were recruited and the interviews conducted through a local support centre for people with sexual convictions. Firstly, this suggests that the participants may have more access to support in the community than many who experience licence conditions, thus findings may not be representative of all individuals' experiences of conditions. Secondly, the proximity to supervisory staff may have moderated some aspects of self-disclosure due to social desirability bias.

Implications for Practice

One of the key implications of this research is to consider how licence conditions are operationalised and administered to individuals. At present, there was general consensus within the data that licence conditions were not individualised and that conditions were things that were done to them rather than with them. There was a desire for more say in the conditions that should apply to them and that the process should be more collaborative. This collaborative approach is likely to improve the perceptions of the conditions, improve compliance (both passively and actively), and enhance the working relationships between probation and the licensee.

There is good evidence for collaborative approaches within the treatment and management of people for sexual offence convictions, e.g., Shingler and Mann (2006) argue that working collaboratively and showing a mutual commitment to the therapeutic process was vital to bolstering engagement in treatment for those with sexual convictions.

It is also linked to legitimacy and procedural justice. Indeed, information gained through a joint, collaborative approach can be more relevant, comprehensive and accurate than that obtained through naïve objectivity, or an adversarial approach dominated by suspicion (Daston, 1992). Procedural justice theory argues that experiencing fair and just procedures (*how* people make decisions and apply policies, rather than *what* the outcome is) leads people to view the law and authority figures as more legitimate and brings greater compliance with, and commitment to obey, the law (Mann et al., 2019. The theory states that the opposite is also true; where procedural justice is lacking, there will be less compliance, both immediately and into the future (Murphy & Tyler, 2008). This is important when managing people in the community, as allowing a collaborative approach, allowing people a chance to "own" or invest in their own rehabilitation, is more aligned with desistance principles (Blagden & Wilson, 2020).

The contention here is that taking a more collaborative approach when setting licence conditions can be seen as an aspect of compassion-focused practice. Compassion has become a major focus for international research in prosocial behaviour in recent times, and the emergence of compassion-focused therapy is beginning to shape forensic practice (see Hocken & Taylor, 2021; Kolts & Gilbert, 2018). We would argue that when someone transitions from prison to the community, it is incumbent on those working with the individual to consider the strengths and 'goods' of that person and to focus on working compassionately with the individual (Blagden & Winder, in press). A turn towards compassion when setting collaborative licence conditions is important as it will assist not just in helping men lead meaningful offence-free lives but also will allow an opportunity for growth and repair.

Conclusion

The findings of this chapter suggest that licence conditions are construed in a critical manner by those under their supervision. Many of the principles of risk monitoring and limiting conditions have been shown in direct contrast to the factors that support desistance. Though some did report positive experiences regarding their interactions with professionals, this was in the void of feeling practically unable to access new personal relationships whilst under licence conditions. Implications from this study implore supervisory professionals to consider personalisation and compassion in the licence process, so that licensees understand the relevance to their reintegration journey but also have a voice in their own rehabilitation.

References

Blagden, N., & Wilson, K. (2020). "We're all the same Here"—Investigating the rehabilitative climate of a re-rolled sexual offender prison: A qualitative longitudinal study. *Sexual Abuse, 32*(6), 727–750.

Blagden, N., & Winder, W. (in press) *Understanding and rehabilitating men with sexual convictions: Theory, intervention, and compassion.* The Oxford Handbook of Criminology Oxford. University Press.

Blagden, N., Winder, B., & Hames, C. (2016). "They treat us like human beings"—Experiencing a therapeutic sex offenders prison: Impact on prisoners and staff and implications for treatment. *International Journal of Offender Therapy and Comparative Criminology, 60*(4), 371–396.

Bohmert, M. N., Duwe, G., & Hipple, N. K. (2018). Evaluating restorative justice circles of support and accountability: Can social support overcome structural barriers? *International Journal of Offender Therapy and Comparative Criminology, 62*(3), 739–758.

Braun, V., & Clarke, V. (2021). Can I use TA? Should I use TA? Should I not use TA? Comparing reflexive thematic analysis and other pattern-based qualitative analytic approaches. *Counselling and Psychotherapy Research, 21*(1), 37–47.

Braun, V., & Clarke, V. (2013). *Successful qualitative research: A practical guide for beginners.* Sage.

Braun, V., Clarke, V., & Weate, P. (2016). Using thematic analysis in sport and exercise research. In B. Smith & A. C. Sparkes (Eds.), *Routledge handbook of qualitative research in sport and exercise* (pp. 191–205). Routledge.

Braun, V., & Clarke, V. (2006). Using thematic analysis in psychology. *Qualitative Research in Psychology, 3*(2), 77–101.

Brown, K., Spencer, J., & Deakin, J. (2007). The reintegration of sex offenders: Barriers and opportunities for employment. *The Howard Journal of Criminal Justice, 46*(1), 32–42. https://doi.org/10.1111/j.1468-2311.2007.00452.x

Burrows, J. (2016). Fear and loathing in the community: Sexual offenders and desistance in a climate of risk and 'extreme othering'. *Moving on from Crime and Substance Use: Transforming Identities*, 153–174.

Casper, J. D., Tyler, T., & Fisher, B. (1988). Procedural justice in felony cases. *Law & Society Review, 22*, 483.

Daston, L. (1992). Objectivity and the escape from perspective. *Social studies of science, 22*(4), 597–618.

de Valk, S., van der Helm, G. P., Beld, M., Schaftenaar, P., Kuiper, C., & Stams, G. J. J. (2015). Does punishment in secure residential youth care work? An overview of the evidence. *Journal of Children's Services.*

Digard, L. (2010). When legitimacy is denied: Offender perceptions of the prison recall system. *Probation Journal, 57*(1), 43–61.

Doekhie, J., Dirkzwager, A., & Nieuwbeerta, P. (2017). Early attempts at desistance from crime: Prisoners' prerelease expectations and their postrelease criminal behavior. *Journal of Offender Rehabilitation, 56*(7), 473–493.

Durling, C. (2006). Never going home: Does it make us safer-does it make sense-sex offenders, residency restrictions, and reforming risk management law. *The Journal of Criminal Law and Criminology, 97*, 317.

Farkas, M. A., & Miller, G. (2007). Reentry and reintegration: Challenges faced by the families of convicted sex offenders. *Federal Sentencing Reporter, 20*(1), 88–92.

Gladfelter, A. S., Lantz, B., & Ruback, R. B. (2018). Beyond ability to pay: Procedural justice and offender compliance with restitution orders. *International Journal of Offender Therapy and Comparative Criminology, 62*(13), 4314–4331.

Guest, G., Bunce, A., & Johnson, L. (2006). How many interviews are enough? An experiment with data saturation and variability. *Field Methods, 18*(1), 59–82.

Harris, P. (2020). 'I think I had to move backwards before I could move forward again': A psychosocial case study exploring the interweaving of desistance from violent offending and professional youth worker identity formation. *Journal of Psychosocial Studies, 13*(2), 193–208.

Her Majesty's Prison and Probation Service (HMPPS). (2021). Licence conditions policy framework. https://assets.publishing.service.gov.uk/government/uploads/system/uploads/attachment_data/file/1010004/licence-conditions-policy-framework.pdf

Hocken, K., & Taylor, J. (2021). Compassion-focused therapy as an intervention for sexual offending. In *Forensic interventions for therapy and rehabilitation* (pp. 189–219). Routledge.

Hudson, K. (2013). *Offending identities*. Willan.

Janicki, M. A. (2006). Better seen than herded: Residency restrictions and global positioning system tracking laws for sex offenders. *BU Pub. Int. LJ, 16*, 285.

Kemshall, H., Dominey, J., & Hilder, S. (2012). Public disclosure: Sex offenders' perceptions of the pilot scheme in England. Compliance, legitimacy and living a "Good Life". *Journal of Sexual Aggression, 18*(3), 311–324.

Knight, M. T., Wykes, T., & Hayward, P. (2003). 'People don't understand': An investigation of stigma in schizophrenia using Interpretative Phenomenological Analysis (IPA). *Journal of Mental Health, 12*(3), 209–222.

Kolts, R., & Gilbert, P. (2018). Understanding and using compassion-focused therapy in forensic settings. *The Wiley Blackwell Handbook of Forensic Neuroscience, 1*, 725–754.

Lasher, M. P., & McGrath, R. J. (2012). The impact of community notification on sex offender reintegration: A quantitative review of the research literature. *International Journal of Offender Therapy and Comparative Criminology, 56*(1), 6–28.

LeBel, T. P., Burnett, R., Maruna, S., & Bushway, S. (2008). The 'Chicken and Egg' of subjective and social factors in desistance from crime. *European Journal of Criminology, 5*(2), 131–159. https://doi.org/10.1177/1477370807087640

LeBel, T. P., Richie, M., & Maruna, S. (2015). Helping others as a response to reconcile a criminal past: The role of the wounded healer in prisoner reentry programs. *Criminal Justice and Behavior, 42*(1), 108–120.

Lomas, J. R. (2021). *House, home, and hope: exploring the accommodation needs and experiences of people with sexual offence convictions*. Nottingham Trent University (United Kingdom).

McAlinden, A. M. (2010). The reintegration of sexual offenders. *Escape Routes: Contemporary Perspectives on Life After Punishment*, 158.

McCarthy, D., & Brunton-Smith, I. (2018). The effect of penal legitimacy on prisoners' postrelease desistance. *Crime & Delinquency, 64*(7), 917–938.

Mann, R. E., Hanson, R. K., & Thornton, D. (2010). Assessing risk for sexual recidivism: Some proposals on the nature of psychologically meaningful risk factors. *Sexual Abuse, 22*(2), 191–217.

Mann, R., Barnett, G., Box, G., Howard, F. F., O'Mara, O., Travers, R., & Wakeling, H. (2019). Rehabilitative culture in prisons for people convicted of sexual offending. *Sexual crime and the experience of imprisonment* (pp. 1–33). Palgrave Macmillan.

Maruna, S., & LeBel, T. P. (2003). Welcome home? Examining the "Reentry Court" concept from a strengths-based perspective. *Western Criminology Review, 4*(2), 91–107.

Maruna, S., & Farrall, S. (2004). Desistance from crime: A theoretical reformulation. *Kolner Zeitschrift fur Soziologie und Sozialpsychologie, 43*, 171–194.

Maruna, S., & Toch, H. (2005). The impact of prison on desistance. In J. Travis & C. Visher (Eds.), *Prisoner reentry and crime in America* (pp. 139–178). Cambridge University Press.

Mews, A., Di Bella, L., & Purver, M. (2017). *Impact evaluation of the prison-based core sex offender treatment programme*. Ministry of Justice.

Ministry of Justice. (2020). *Offender management statistics, probation (2020)*. Ministry of Justice.

Murphy, K., & Tyler, T. (2008). Procedural justice and compliance behaviour: The mediating role of emotions. *European Journal of Social Psychology, 38*(4), 652–668.

National Offender Management Service. (2015). *Licence conditions, licences and licence and supervision notices*. Ministry of Justice. Retrieved from: https://www.justice.gov.uk/downloads/offenders/psipso/psi-2015/psi-12-2015-licences-conditions-supervision.pdf

Nugent, B., & Schinkel, M. (2016). The pains of desistance. *Criminology & Criminal Justice, 16*(5), 568–584.

Oliver, D. G., Serovich, J. M., & Mason, T. L. (2005). Constraints and opportunities with interview transcription: Towards reflection in qualitative research. *Social Forces, 84*(2), 1273–1289.

Perrin, C., & Blagden, N. (2014). Accumulating meaning, purpose and opportunities to change 'drip by drip': The impact of being a listener in prison. *Psychology, Crime & Law, 20*(9), 902–920.

Perrin, C., Blagden, N., Winder, B., & Dillon, G. (2018). "It's sort of reaffirmed to me that I'm not a monster, I'm not a terrible person": Sex offenders' movements toward desistance via peer-support roles in prison. *Sexual Abuse, 30*(7), 759–780.

Reid, K., Flowers, P., & Larkin, M. (2005). *Exploring lived experience. The psychologist*. BPS.

Richards, K. (2022). Circles of support and accountability: The role of social relations in core member desistance. *International Journal of Offender Therapy and Comparative Criminology, 66*(10–11), 1071–1092.

Rydberg, J. (2018). Employment and housing challenges experienced by sex offenders during reentry on parole. *Corrections, 3*(1), 15–37.

Shingler, J., & Mann, R. E. (2006). Collaboration in clinical work with sexual offenders: Treatment and risk assessment. *Sexual Offender Treatment: Controversial Issues*, 225–239.

Skrapec, C. A. (2001). Phenomenology and serial murder: Asking different questions. *Homicide Studies, 5*(1), 46–63.

Sturge, G. (2021). *UK Prison population statistics*. House of Commons Library. Retrieved 16 September, 2022, from https://commonslibrary.parliament.uk/research-briefings/sn04334/

Tewksbury, R., & Copes, H. (2013). Incarcerated sex offenders' expectations for reentry. *The Prison Journal, 93*(1), 102–122.

Tewksbury, R., & Mustaine, E. E. (2009). Stress and collateral consequences for registered sex offenders. *Journal of Public Management & Social Policy, 15*(2).

Tewksbury, R., & Lees, M. (2006). Perceptions of sex offender registration: Collateral consequences and community experiences. *Sociological Spectrum, 26*(3), 309–334.

Tolson, D., & Klein, J. (2015). Registration, residency restrictions, and community notification: A social capital perspective on the isolation of registered sex offenders in our communities. *Journal of Human Behavior in the Social Environment, 25*(5), 375–390.

Tovey, L., Winder, B., & Blagden, N. (2022). 'It's ok if you were in for robbery or murder, but sex offending, that'sa no': a qualitative analysis of the experiences of 12 men with sexual convictions seeking employment. *Psychology, Crime & Law*, 1–24.

Ugwudike, P. (2010). Compliance with community penalties: The importance of interactional dynamics. In McNeill F, Raynor, P. and Trotter C (Eds) *Offender Supervision: New directions in theory, research and practice*. Willan.

Waddell, G., & Burton, A. (2016). Is work good for your health and well-being? An independent review. Department for Work and Pensions.

Wagner, C. (2010). The good left undone: How to stop sex offender laws from causing unnecessary harm at the expense of effectiveness. *Am. J. Crim. L.*, 38, 263.

Ward, T. (2010). The good lives model of offender rehabilitation: Basic assumptions, etiological commitments, and practice implications. In F.

McNeill, P. Raynor, & C. Trotter (Eds.), *Offender supervision: New directions in theory, research and practice*. Willan Publishing

Ward, T. (2017). Prediction and agency: The role of protective factors in correctional rehabilitation and desistance. *Aggression and Violent Behavior, 32*, 19–28.

Weaver, B. (2015). *Offending and desistance: The importance of social relations*. Routledge.

Weaver, B., & McNeill, F. (2015). Lifelines: Desistance, social relations, and reciprocity. *Criminal Justice and Behavior, 42*(1), 95–107.

Williams, F., & Bailey, J. (2019). *Working with people with sexual convictions in HMPPS*. Paper presented at Symposium on key issues in the imprisonment of people convicted of sex offences, University of Cambridge.

Willis, G. M. (2018). Why call someone by what we don't want them to be? The ethics of labeling in forensic/correctional psychology. *Psychology, Crime & Law, 24*(7), 727–743.

Winder, B. (2022). *Reducing reoffending and increasing community (re)integration: Effective practice when people have a sexual conviction.* Clinks Evidence Library. https://www.clinks.org/publication/reducing-reoffending-and-increasing-community-reintegration-effective-practice-when-people-have-a-sexual-conviction

Zevitz, R. G., & Farkas, M. A. (2000). Sex offender community notification: Managing high risk criminals or exacting further vengeance? *Behavioral Sciences & the Law, 18*(2–3), 375–391.

8

Exploring Trauma-Informed Care in a High-Intensity Programme

David S. Prescott and Jill Levenson

Introduction

This chapter explores trauma-informed care in a high-intensity treatment programme. Trauma-informed care (TIC) is a service delivery paradigm that incorporates evidence about the prevalence and impact of early trauma on behaviour across the lifespan. Trauma-informed treatment integrates data about trauma into practices that promote respect for clients, encourage self-determination, and facilitate the rebuilding of healthy relational skills to replace maladaptive survival skills that developed in the face of childhood adversities. TIC occurs in a safe and client-centred environment in which service providers conceptualise

D. S. Prescott (✉)
Safer Society Foundation, Brandon, VT, USA
e-mail: davidprescottvt@gmail.com

J. Levenson
College of Health Professions and Medical Sciences, School of Mental Health & Social Work, Barry University, Miami Shores, FL, USA

N. Blagden et al. (eds.), *Sexual Crime and Community Reintegration*, Sexual Crime,
https://doi.org/10.1007/978-3-032-06034-1_8

presenting problems within the context of a trauma-informed treatment framework.

Dozens of studies demonstrate significant correlations between higher rates of childhood adversities and increased risk for a range of medical and behavioural disorders, including chemical dependency, physical diseases, and behavioural health disorders (e.g., Anda et al., 2006; Douglas et al., 2010; Edwards et al., 2003; Weiss & Wagner, 1998). In addition, children who are exposed to multiple, inter-related traumas and poly-victimisation often develop complex trauma reactions (Cloitre et al., 2009; Finkelhor et al., 2011; Maschi et al., 2013; Streeck-Fischer & van der Kolk, 2000). In this chapter, we will use a case example to conceptualise the client's problems by understanding how early trauma impacted his thoughts, feelings, behaviour, and relational style. Trauma-informed sex-offending treatment (SOTX) addresses dynamic risk by recognising trauma symptoms, reducing barriers to building good lives, and promoting adaptive coping, creating conditions that reduce the risk for recidivism (Harris & Levenson, 2020; Levenson et al., 2017).

Jason[1]

The following case example illustrates the complexities of understanding and helping individuals who have sexually abused as they enter and participate in therapy, and then gradually reintegrate closer to and into the community.

Biography and Background Information

There is a great deal of Jason's life that he simply cannot recall. Several possible reasons might explain gaps in memory, including the fact that so much of it was traumatic, and he has a strong aversion to trying to remember it. Another reason is that he got into a fight with another male as a teenager and suffered a traumatic brain injury. This injury further aggravated several aspects of his functioning that precluded his living independently in the community.

From what can be pieced together, Jason had a painful childhood, marked by persistent physical and sexual abuse by his paternal uncle over the course of several years. Jason's parents were Native American and gave him up for adoption at the time of his birth. His adoptive parents went on to experience multiple misfortunes. Jason's father began drinking heavily and had a severe substance use disorder throughout much of his childhood, eventually disappearing from his life. His mother attempted to compensate for this condition by working two jobs and doing her best to raise her children. There are questions regarding to what extent she may have turned a blind eye to some knowledge of the abuse Jason suffered due to her financial and emotional dependence on his father's side of the family. Although it may seem deeply unwise to outsiders, she may have perceived the dire needs of her circumstances to be so urgent that the abuse of Jason went unrecognised and/or unaddressed. Rather, living and surviving were her highest priorities.

Jason experienced considerable covert racism from an early age. The other children in his predominantly white school and community did not socialise with him. Their parents and his teachers often spoke to him differently than they did the white children in the school. As he grew older, the adults in his community were far less kind to him than to others, even though they had some awareness of the circumstances in which he was growing up. Over time, Jason would come to understand himself as coming from a cultural heritage destroyed by people who looked like those living in his neighbourhood. He was very aware that in many other situations, Native American tribes had successfully blocked adoptions by people outside of their culture, although this had not happened in his case. However, it caused him to wonder how his life might have been different if he had stayed closer to people with his heritage.

Although he was subject to very little overt racist taunting by others, he grew up experiencing subtle micro-aggressions from others. For example, he noticed that his teachers always took a more formal tone with him and kept themselves at a greater distance from him in the classroom than they did with other students. When it was necessary to re-assign seats at school or community activities, Jason was always the first to be told to move. Likewise, business owners and employees often looked at him with a level of suspicion that they did not display toward the other kids in his community. He came to view these small experiences, which happened nearly every day, as a kind of echo of the experiences of his ancestors, who were moved from their homeland to foreign areas, discriminated against, and in many instances, murdered outright.

While Jason does not remember much of his childhood, he has a clear memory of one day in particular. It was in the wintertime. His school was closed due to severe weather, and his mother stayed home to care for him. This provided her with an opportunity to catch up on domestic chores, starting with taking their dirty clothes to a public laundry. In a complicated situation involving a broken dryer, Jason ended up needing to wear a pair of women's underwear during the visit and while they walked home. Although this lasted only a brief time, Jason never forgot two features of that day. The first was the sense of connection he felt with his mother; they rarely had the opportunity to spend much time together without others, including Jason's father and uncle. This was a rare mother-son moment and it felt good. The second feature was that Jason took note of the feeling of the underwear on his skin. It felt much better than any of his other clothes and left a profound impression on him.

Putting these events together in his mind, Jason developed a strong sense of intrinsic gratification from wearing women's underwear. He described it as a way to get in touch with a part of him that he could not experience very often. When dressed in women's underwear, Jason felt connected to a side of himself that felt more loving, softer, and warm than men seemed allowed to be. He also felt more lovable. However, for others, it would present a range of challenges for working with him that are described below.

As he entered adulthood, Jason experienced multiple internal and external barriers to living independently and becoming who he wanted to be. His adverse early life experiences left him deeply confused about the nature of interpersonal relationships. He had few bonds with others, and each of his relationships (with his mother, father, uncle, and others) had all been violated in one way or another. His attachments to others were primarily disorganised, meaning he experienced them as sources of danger interspersed with occasional nurturing. He developed a core cognitive schema of himself as a broken and unlovable person who needs to find and take advantage of momentary states of happiness and pleasure wherever he can find them because no one will ever give these experiences to him. To others, he would appear an opportunist whose impulsive actions could seem to come out of the blue. To Jason, he was simply taking advantage of whatever circumstance he could to have fun against the bleak backdrop of a world he would never fully understand, much less participate in. His most common behaviours involved attempts to find drugs and alcohol, and to make unwanted advances toward teenage girls and adult women in an attempt to persuade them to have sex with him.

By the time he was 18 years old, Jason had developed a substance use disorder of his own. A lifetime of adversity, unhealthy attachments, a cognitive style that kept him constantly focused on the present moment (and on the future only rarely), and a self-regulation style adapted to survive a hostile world had left him with serious challenges. Despite adapting to the traumatic world of his past, Jason struggled to develop a self-determined lifestyle for his future.

At the age of 19, Jason was arrested on drug charges. With no one to provide him with meaningful assistance, he spent just under a year in a county jail in his home state in the US. Upon his discharge, he lived briefly with his mother and ended up homeless for several weeks. His time in the jail had itself been traumatic, with each day seeming to confirm his lifelong belief that the world is an inherently dangerous, out-of-control place where one wakes up every morning to the possibility of needing to protect oneself from harm. He received virtually no case-management services until the very end of his sentence due to an overburdened social work department. He was able to engage in some psychoeducational classes, but these were of little help to him. By the time he re-entered the community, he had few options for housing or employment. His criminal record made these challenges even more daunting.

Jason's recollections of his time are complicated. To anyone unfamiliar with trauma, he would appear to minimise or make light of his time in jail. Sometimes, he might say he doesn't remember much of it. However, closer examination would show that Jason lacks the verbal skills to put words to his experience. Further, his earlier traumatic brain injury might also play a role in his difficulties remembering it. However, it is important to remember that much of his downplayed response may itself be a trauma reaction to his experiences. It is important to consider that the diagnostic criteria of PTSD include an inability to recall key features of traumatic events and avoidance of trauma-related stimuli after the event, such as trauma-related thoughts and feelings or external reminders. In Jason's case, he also experienced decreased interest in activities, difficulties concentrating and sleeping, hypervigilance, and an increased tendency to engage in risky and/or destructive behaviours. In other words, trauma symptoms are often disguised as presenting problems.

Upon his return to his mother's apartment, she made calls to nearby agencies and managed to find a case manager for Jason. The case manager rightly sought out a collection of assessments, which coalesced around findings that Jason could not live independently due to his traumatic brain injury. Under the statutes and policies of his home state, this injury entitled him to a range of services. Accordingly, he was placed in a supervised apartment with two other young men and staff members who would take different shifts providing supervision and community access.

Unfortunately, this arrangement lasted only briefly. Jason attempted to sexually assault an employee of a nearby coffee shop that he visited often. Jason said that he found himself attracted to her. When becoming aroused in the past, Jason would wear women's underwear and/or masturbate, with or without pornography. Now, for essentially the first time, he found himself in close quarters with an adult female to whom he felt attracted. The circumstances were such that the staff of his programme that was supposed to supervise him left him alone briefly to make a phone call. While he was aware that the rules of his programme prevented sexual contact, he also had a long history of opportunistic behaviours and, seeing an opportunity, moved closer and forced himself on her. She screamed, and the staff member came back and separated them. The employee quickly called the police while the staff member attempted to physically restrain Jason. At the moment the police arrived, Jason took off all of his clothes, waved the female undergarment he was wearing around his head, and invited the arresting officers to take him away.

Jason was hospitalised briefly and placed into a much larger, inpatient, high-intensity treatment programme, where he would reside for several years. This followed the legal determination that Jason required a court-appointed legal guardian and was not competent to stand trial. As a result, Jason has faced considerable difficulties as he has tried to move forward with his life.

Treatment Considerations of the Past and Present

In the authors' experience, traditional treatments of Jason would have focused on two primary areas: the sexual assault on the staff and his fetishistic attraction and behaviours related to women's underwear. Depending on the programme, treatment for his substance use disorder would occur separately from his offending behaviour and fetishistic disorder, which might also be viewed as separate but related. In the US, funding streams for different kinds of treatments might necessitate that they each occur simultaneously, with little communication between the providers who each treat only their focus areas; no one would be responsible for developing a holistic understanding of Jason. Relying primarily on a cognitive-behavioural framework, his treatment would focus on:

- reducing impulsivity
- addressing factors in Jason's clinical presentation that interfere with meaningful participation in treatment
- disclosing offending behaviour
- understanding the antecedents of his offending: patterns of thoughts, attitudes, beliefs, feelings, situations, and behaviours that preceded offending
- developing and enacting skills for managing these antecedents
- developing plans for avoiding risky situations in the future and managing those that would come up

The central feature of this traditional sex-offending treatment (SOTX) approach would lie in exploring everything that had gone wrong in Jason's life without a strong focus on what had gone right and what he might do to build on his strengths. At no point would his cultural heritage (and cultural trauma) or the role of early adversity and trauma be addressed. Rather than help Jason understand how he came to be who he is, and explore who he could be in the future, treatment would focus primarily on trying to fix his behaviours and his many disorders. In many cases, the therapist would adopt the stance of being in a one-up position, someone administering treatment from an implicit position of moral authority.

Depending on circumstances, those providing treatment for Jason's sexually problematic behaviour would likely view his traumatic brain injury as a risk factor for offending, given its contribution to impulsive behaviour. He would be given medication and psychoeducation to ameliorate its effects. However, this treatment would not explore how he might use this disability in the service of helping others (for example, helping others who have had similar experiences). In many situations, community-based placements and programmes with expertise in treating individuals with traumatic brain injuries might not want to work with Jason due to his offending behaviours. On the other hand, many programmes focusing on offending would be concerned that they do not have enough specialisation to provide treatment based on a deep understanding of traumatic brain injury. Programmes that can address a person's complex needs can be difficult to find.

Current, Trauma-Informed Care Approaches

In contrast, Jason's programme today actively provides trauma-informed care, and the clinical staff understands the intersection between past trauma and current patterns of thoughts and behaviour. Trauma can change the affective and cognitive processing by which individuals attach meaning to their experiences by which we "establish a coherence of functioning as a thinking, feeling human being" (Rutter & Sroufe, 2000, p. 265). Our experiences inform our expectations, and our expectations are often projected into the interpretation of our current circumstances and interactions with others. For example, children who experience chronic trauma or deficient parenting in childhood can develop maladaptive cognitive schemas, disorganised attachment patterns, emotional dysregulation, and Cluster B personality traits (Chakhssi et al., 2013; Loper et al., 2008; Young et al., 2003).

Adverse childhood experiences produce neurobiological changes in the brain that can impact development across the lifespan. Toxic stress leads to an over-production of hormones associated with survival responses (fight or flight), preventing neural connections in the brain, and impeding cognitive processing and self-regulation capacities into adulthood (Alink et al., 2012; Creeden, 2009; Finkelhor & Kendall-Tackett, 1997; SAMHSA, 2014; Streeck-Fischer & van der Kolk, 2000; van der Kolk, 2006). Research has indicated that the multiplicity, frequency, and chronicity of early adversity creates what has become known as a complex trauma response, which manifests in maladaptive coping strategies, mental health symptoms, personality pathology, and relational problems. (Cloitre et al., 2009; Herman, 1992, 1997; van der Kolk, 2014).

On the other hand, focusing on strengths and resilience is crucial to healing and recovery. Therefore, Jason's clinical team, as well as the line staff, also get training in using the good lives model and motivational interviewing as overarching approaches. Importantly, treatment begins with an assessment conducted by an outside evaluator, assigned by the referring agency and not the programme itself. The task of the evaluator is first to address areas of risk, need, and responsivity (Andrews & Bonta, 2010). Next, the evaluator explores the biopsychosocial characteristics of the client and makes recommendations. This involves conceptualising early adversity and trauma in the client's life. Along the way, client strengths and protective factors are also assessed. The outcome is a report that helps to guide holistic, culturally sensitive, strengths-based, and trauma-informed interventions. From there, the treatment program can establish methods to pursue two goals:

- building Jason's capacities in reducing and managing risk

- developing a balanced and self-determined lifestyle in which Jason views offending as unnecessary and undesirable

Within the programme, the administrators continually emphasise to the line staff that many of the less pleasant aspects of Jason's interactive style (for example, opposition to and defiance of rules) are themselves symptomatic of his lived experiences from an earlier age. Diagnostically, it is important to consider that the dysregulated behaviour (impulsivity, oppositional defiance, addiction, aggression) often traum (intrusive or ruminating thoughts, avoidance, detachment, recklessness, risk-taking, difficulties concentrating, hyperarousal, and negative thoughts or moods). Indeed, personality disorders, described as pervasive patterns of thoughts, feelings, and behaviours, typically develop in early life as maladaptive responses to interpersonal experiences, many of which are traumagenic. Simply put, complex trauma results from the accumulation of multiple and chronic traumatisation, including developmental and relational trauma, cultural and intergenerational trauma, and trauma related to arrest, conviction, incarceration, and registration. Trauma produces dysregulated emotions, feelings, thoughts, defences, reactions, responses, and behaviour, and interferes with cognitive processing and memory consolidation (van der Kolk, 2005).

The staff are reminded that the most effective staff are warm, empathic, rewarding, and guide clients through the routines of the day. The administrators will also remind staff that if they find themselves becoming harsh or confrontational, they are emulating aspects of the environment in which Jason grew up. This ongoing education serves to help the staff be more patient and understanding without sacrificing adherence to the routines of the programme, which themselves provide a sense of safety and predictability. By shifting the paradigm to view client problems as symptoms of patterns by which they responded to traumagenic experiences, we can begin to contextualise treatment not as "what's wrong with you?" but as "what happened to you?".

Treatment involves many of the same aspects as treatments of the past, including discussion of offending behaviours and their antecedents. However, it also emphasises a deep understanding and appreciation of

Jason's earliest experiences and how they came to influence him. In addition, it explores how he came to be the person he is, as well as how he came to offend. Jason is, of course, clear that he alone is responsible for his behaviour. However, understanding and appreciating how he came to cause harm to others enables him not only to develop the skills to prevent further harm, but also to develop a more profound sense of self-compassion.

Corrective Experiences Through TIC

A brief formulation of Jason's case is that his early life experiences and cultural adversity wreaked havoc on his capacity to form secure attachments to others. While others were learning how to expand their relationship capacities outward towards schoolmates and community members, Jason was learning that others could be dangerous and hurtful. Jason developed implicit theories about the world as a dangerous and out-of-control place where men are hostile, and women are weak, and often exploitable. These experiences were so deeply ingrained in his consciousness from an early age that it was easy to dismiss any evidence that this worldview was wrong (this might be thought of as a survival-based confirmation bias). Because of this, he viewed the world as potentially dangerous at every turn and spent his energies scanning every circumstance for evidence of threats. As a result, his behaviour towards others often appeared impulsive, and his coping style focused on fighting back rather than seeking solutions.

Early childhood trauma compels hyper-arousal, by which the brain and body constantly scan the environment for danger; Jason has been primed to expect threats and to respond accordingly. As a result, some people develop a sense of learned helplessness, and others take on a stance of proactive aggression. Their self-protective survival mechanisms may manifest in aggressive, combative, or provocative behaviour (fight response), avoidance of intimacy or tendency to self-medicate with substances, food, or compulsive behaviour (flight response), or passive compliance, dependency, and difficulties setting boundaries (freeze response). Therapy helps Jason to engage in better reality testing and self-correction, and to devise new strategies to face vulnerability with healthy coping, intact boundaries, and an expanded repertoire of problem-solving skills.

Unsurprisingly, earlier risk assessments noted relationship instability, social isolation, attitudes tolerant of violence and criminality, impulsivity, and a dysfunctional coping style. However, these dynamic risk factors were themselves the logical results of earlier experiences. While they certainly contributed to risk, these factors would be better thought of as the processes by which Jason had managed potential existential threats. Clearly, these responses to potential danger were no longer serving him. However, these same risk factors functioned as barriers to Jason's leading the best life that he could. He lacked the capacities he needed to form meaningful relationships and to experience himself as competent within these relationships. His surviving threats by responding impulsively at every turn came with the cost of not being able to achieve autonomous functioning in most life domains. It also prevented him from becoming genuinely competent in anything. Finally, he had few ways of experiencing momentary states of happiness and pleasure outside of his interest in women's underwear. When treatment focused on building his competencies in these areas, Jason gradually felt more at ease with himself and others.

Jason's therapist, Lorraine, took a compassionate approach to providing therapy. She reminded herself before every session that, as a common expression has it, Jason had long been asked what was wrong with him instead of what had happened to him. Lorraine reminded herself to make sure that she also asked what was right with Jason and what strengths were there for her to work with. Further, as per recent research, Lorraine sought to demonstrate warmth and empathy and provide direction to treatment, along with many statements of validation and affirmation. This approach created a sense of safety and trust in the therapeutic environment, bolstered by an empowering collaboration by which she and Jason can work together to identify and prioritise treatment goals. As Jason learns to pay more attention to his physiological reactions to stressors and begins to understand how they are connected to protective responses to the vulnerability that was adaptive in his childhood, he can view his present reality in its current context and make better coping choices.

There are many ways that the very interventions designed to help can re-traumatise clients. When clients act out, we often resort to limit-setting and punitive consequences. We want to teach a lesson to provide deterrence for others. We label, pathologise, diagnose, and medicate. We worry about our own liability. However, these responses rarely provide the re-conditioning we hope to stimulate; indeed, they often reinforce clients' feelings of mistrust and vulnerability. Traumatic reenactment occurs when a client's attempts to cope or adapt lead to self-fulfilling prophecies of failure—fortifying their anger, depression, or inflexible styles of thinking and interacting with others. Instead, Jason's therapist tries to understand why he is reacting a certain way, validates feelings, asks questions with curiosity and compassion, and tries to connect with him in a respectful encounter that provides something different from what the client expects. We might ask ourselves, "why does the client need to do this?".

Adjunctive Treatments

As in the past, cognitive-behavioural principles are at the centre of Jason's treatment. However, adjunctive treatments are also provided so that Jason can better access the full range of services available to him. Jason's trauma and adversity resulted in underdeveloped abilities to understand his cognitions and behaviours. While most people have some ability to notice their thoughts as they have them and to act accordingly, Jason has not had these abilities throughout much of his life. To survive his historical environments and circumstances, Jason constantly looked outside himself to spot any evidence of imminent threats to his physical and psychological integrity. Despite the many challenges he faced in his life, he was highly skilled at assessing others' states and could often surmise which staff members had arrived at work tired or in negative moods.

While Jason had abilities in "reading" others, they had come at the cost of his ability to understand himself. In fact, across much of his life, he had appeared barely able to understand that he had thoughts. His thoughts and behaviours had become so automatic as a result of his early-life experiences that asking him to change them was nearly meaningless until he could understand that:

- thoughts do not necessarily reflect reality;
- different people have different thoughts;
- people can notice how they think;
- people can consider and give thought to how they think (known as meta-cognition);
- people make decisions about how they think;
- people can change the way they think;
- by changing how they think, people can create possibilities for behaving differently;
- people can use thinking skills to take effective actions with their behaviours.

Psychologist Jeffrey Young and his colleagues created a list of what they called "early maladaptive schemas," which are flawed patterns of thinking that develop based on experiences growing up. Early maladaptive schemas (EMS) are ways we are programmed to think about ourselves, others, and the world around us based on our early experiences and family life. EMS impact how we think about things that happen to us: how we interpret events, our expectations of others, and what our experiences seem to mean to us. These things we believe about the world become our fundamental beliefs. So, work on cognitive distortions—not only about offense behaviour, but also about worldview and EMS—are essential in devising more healthy and adaptive cognitive coping skills (Levenson & Morin, 2021).

While professionals could see that Jason's thought patterns and behaviours reflected core cognitive schemas about the world and entrenched thinking patterns, Jason would require considerable work to develop the metacognitive skills that would lead to more effective action. Before Jason could change his thoughts and behaviours, he would first have to be able to observe them. To this end, the staff members of his programme implemented a wide range of activities, including meditation, trauma-sensitive yoga, and biofeedback.

Although each of these adjunctive treatments has been the subject of numerous books and research studies, their application to Jason's treatment was reasonably straightforward. Jason would meditate with others for ten minutes twice daily. The staff members would play recordings of ocean waves and the like for their soothing properties, but Jason knew that the idea was to focus on his breath and notice thoughts as they arose. Jason initially liked meditation's obvious benefits of stress reduction and calming. However, he found his ability to notice thoughts and impulses most helpful over time. For example, when others in the programme tried to make him angry, Jason subtly shifted from experiencing anger instantly to noticing the emergence of anger and slowly becoming able to make decisions based on what he noticed.

Likewise, Jason participated in yoga specifically designed for people who have been traumatised. This form of yoga involves gentle stretching and aligning breath with movement. It involves not making judgments about the movements but instead simply noticing the experience, particularly the physical sensations. Jason learned that turning one's attention inward to physical sensations is known as interoception, which is closely related to proprioception, or the noticing of one's body in movement through space. For Jason, this meant that not only was he becoming more aware of the physical signs of anger and anxiety, but that he could make decisions accordingly. As he watched how his thoughts and sensations unfolded, he gradually learned that he could allow difficult thoughts and emotions to simply pass from his experience rather than feeling compelled to act on them immediately.

Finally, his biofeedback taught him how he could intentionally keep himself calm and feel more in control of his various mental states. Then, using a computer-based programme, he could watch his progress in real-time. Over time, he developed not only skills at self-observation, but self-confidence in his abilities to notice and observe his thoughts and behaviours, as well as to change them. In all, Jason made considerable progress in treatment, gradually becoming better able to manage his behaviours and solve problems in daily life as they arose.

Barriers to Full Community Reintegration

Like most people in his situation, Jason had tremendous difficulty in his attempts to integrate back into the community. In many ways, he was lucky because he did not have to repeat his experiences of first being incarcerated. During that episode, he was simply released with few resources and ended up homeless for a while. Now that he was involved in a community-based programme, Jason had others who realised the importance of helping him to develop the skills for a productive, offence-free, community life.

Challenges from the outset included internal obstacles to community access. Jason's case-management team would not soon forget his attempt to sexually assault a staff member. This meant that he would need to make significant changes to his behaviour in treatment. Although he understood this, he also longed for the ability to experience life outside the confines of a programme. He found himself in a classic bind experienced by clients in treatment programmes: His treatment team wanted to see him demonstrate independent-living skills in an environment that is very different from actual independent living. When Jason would comment on this ("How can I show you I'm ready to be there if I'm still here almost all the time?"), the team would respond by asking him to demonstrate a high level of self-regulation throughout his day.

However, even the best inpatient treatment programmes can be destabilising environments in which it is challenging to demonstrate self-regulation. As Jason would note, his attempts to show that he is ready to be in the community were often in an environment in which he was not allowed to cook his food, dial the phone to call his mother, or possess items (such as a computer) that he would have access to in the community. The net result was that Jason often had considerable difficulty demonstrating the skills his team requested in daily life, and few other options for demonstrating other relevant skills, such as responsible use of the Internet. Ironically, Jason's treatment team could see that his circumstances were difficult but offered no clearer path toward building trust.

Gradually, Jason demonstrated his self-regulation skills well enough that he was allowed to volunteer at an animal shelter in the community.

This was a coveted position within the programme, and he was glad to have it. Over the course of ten hours a week, he would help to clean the shelter and take care of the animals. The programme administration had formed a good working relationship with the shelter staff over the years. Jason experienced considerable success in this environment. The main challenge he experienced was the jealous taunting he sometimes received from the others in his programme.

In this environment, Jason started to see the benefits of his treatment. Where once he might have warily viewed some of the animals as potentially out to bite or scratch him, his newfound abilities to observe and regulate his mood were apparent in this setting. Jason soon found that developing the right mindset for handling these animals was not only possible, but that he could do it. The shelter had become something akin to a biofeedback device: when he was well-regulated, the animals generally responded accordingly. Perhaps just as importantly, the more Jason was able to take effective action in maintaining the right attitude, the more completely he appreciated the animals in his care. He began to understand their experiences and care about them very much.

Jason's experiences in this shelter offer a caution about how professionals understand the extant research into what works in treatment. For example, past literature has emphasised the importance of risk, need, and responsivity principles. Indeed, many scholars would argue that animal shelters or similar volunteer work would have no place in models of correctional treatment (the first author once heard a research criminologist scoff about how "the only outcome of gardening programmes is vegetables"). While it is true that one should not use volunteer work with the expectation that it will directly reduce recidivism risk, with the right clients such activities can act as venues for practising skills and demonstrating change. In Jason's case, he reviewed and explored his experiences in depth with his therapist in the programme. As a result, he could finally examine patterns of thoughts and behaviours that were meaningful and relevant to him in his present circumstances. As a result, he began to work more diligently in managing his experiences in the here and now. Although his work exploring historical behaviours had been necessary, he was grateful for the opportunity to show others in his life that he was no longer the person he had been.

From there, Jason had gained enough credibility with his treatment team that he began to look for a job. This proved far more difficult, as the local businesses in the community were sceptical about hiring people on probation and in care. Further, in Jason's situation, his case manager and treatment team felt it was essential to have a staff member supervise him for several months to ensure his safety and that of others. Although more restrictive than Jason might have liked, the prospect of actual employment made it worthwhile. The programme had staff whose job involved calling around to local businesses, and eventually, a small number became willing to work with the programme in an exploratory fashion.

Jason got a job working for a construction materials supply company. He loaded and unloaded trucks. He worked with others who would not tolerate anything less than giving one's best effort. Although initially nervous, Jason quickly became attuned to the job. Once again, he found that the skills he had developed in treatment could help him. In addition, Jason found that the demanding and physical nature of the work provided him with the opportunity to focus on nothing else. As he explained, "When I am working there, nothing else matters. My past doesn't haunt me, and my future doesn't concern me. And I feel good. My body feels good doing it."

Jason's employment also presented challenges, but the staff member supervising from a distance never had to intervene. Jason quickly noticed that he sometimes questioned the intent of others or seemed to take things personally. Historically, these would have been precisely the times he was at the highest risk of becoming enraged or quitting and going someplace to get high. He developed a motto: "When anything happens, just talk to someone." Over time, he began to see real successes, not only in negotiating the stress of daily life, but in gaining the respect and confidence of others around him.

Jason would return to the programme and review even the slightest moments of discomfort by reviewing the thoughts and feelings he experienced, how they were connected to historical, trauma-based cognitive schemas, and the decisions and behaviours that he engaged in as a result. As with the animal shelter, he continued to find considerable meaning both in his work and in his therapeutic efforts at improving himself in the workplace. His case manager and treatment team all took notice.

Lessons

What are we to learn from Jason's efforts at reintegration? Perhaps one of the most important lessons is how different Jason's experience was from many others. Individuals often leave incarceration suddenly and with almost no support or resources. Indeed, this had happened to Jason earlier. The ability to access the community that one is trying to reintegrate into (or perhaps integrate into for the first time) has become less common in many jurisdictions where individuals are held until the end of their sentences and then released without meaningful safeguards. Understanding clients through a trauma-informed lens brings the knowledge that expecting many clients to reintegrate without support or resources is tantamount to increasing their risk for failure, sometimes insurmountably. While the general public often wants to see people who have abused others be kept out of the community for as long as possible, waiting until the last minute of a person's sentence is generally counterproductive.

We also see from Jason's case that as helpful as cognitive-behavioural therapy utilising risk, need, and responsivity principles can be, adopting a trauma-informed perspective can substantially enhance the benefits of these approaches. Using adjunctive therapeutic activities and integrating work and volunteer activities into a broader suite of treatment programming can help clients make deeper and more meaningful changes to their lives.

Conclusion

Rather than thinking of treatment as being a series of conversations and activities that reduce the effects of specific risk factors, trauma-informed care can help clients and professionals understand the origins of those risk factors as well as their narrative qualities. Questions that professionals might ask include:

- What is the story, or narrative, of this risk factor in the client's life?

- How have adverse events, including cultural adversity and prior experiences with the legal system, contributed to the role of this risk factor in the client's life?
- How has the client come to view the world? What kinds of cognitive schemas does the client have about the world? How have these schemas contributed to risk?
- What kind of cognitive schemas has the client developed about themselves? As someone who is broken? Who has to fight back against others? Who has to take what he wants in an unkind world? How have these schemas contributed to risk?
- How does the client understand the role of this risk factor in their life?
- To what extent are they able to observe and work with this risk factor?
- In what ways has this risk factor represented an obstacle to living a better life?

Treatment should address dynamic risk factors and criminogenic needs for each individual, including self-regulation problems, intimacy deficits, deviant sexual interests, criminality, lack of empathy, and co-morbid conditions such as substance abuse, anxiety, depression, or PTSD (Andrews & Bonta, 2010, 2017; Levenson et al., 2017; Yates et al., 2010). Professionals should be aware that current stressors and constant anxiety can re-activate maladaptive coping styles and increase dynamic risk (Harris & Levenson, 2020). Under such circumstances, dysregulation can undermine protective factors such as stable housing, prosocial relationships, meaningful employment, and appropriate leisure activities—all of which increase one's sense of self-esteem, social belonging, and self-actualisation and protect against reoffending (de Vries Robbé et al., 2015; Kruttschnitt et al., 2000).

Notes

1. Jason is a fictitious case based on the authors' experiences over decades of providing and supervising treatment as well as consulting to therapists and agencies. The elements of this case have been

woven together from common themes seen in in inpatient practice. Any resemblance to any individual is purely coincidental.

References

Alink, L. R., Cicchetti, D., Kim, J., & Rogosch, F. A. (2012). Longitudinal associations among child maltreatment, social functioning, and cortisol regulation. *Developmental Psychology, 48*(1), 224–236. https://doi.org/10.1037/a0024892

Anda, R. F., Felitti, V. J., Bremner, J. D., Walker, J. D., Whitfield, C., Perry, B. D., Dube, S. R., & Giles, W. H. (2006). The enduring effects of abuse and related adverse experiences in childhood. *European Archives of Psychiatry and Clinical Neuroscience, 256*, 174–186. https://doi.org/10.1007/s00406-005-0624-4

Andrews, D. A., & Bonta, J. (2010). Rehabilitating criminal justice policy and practice. *Psychology, Public Policy, and Law, 16*(1), 39–55. https://doi.org/10.1037/a0018362

Andrews, D. A., & Bonta, J. (2017). *The psychology of criminal conduct* (4th ed.). Anderson Publishing.

Chakhssi, F., Ruiter, C., & Bernstein, D. P. (2013). Early maladaptive cognitive schemas in child sexual offenders compared with sexual offenders against adults and nonsexual violent offenders: An exploratory study. *The Journal of Sexual Medicine, 10*(9), 2201–2210. https://doi.org/10.1111/jsm.12171

Cloitre, M., Stolbach, B. C., Herman, J. L., Kolk, B. v. d., Pynoos, R., Wang, J., & Petkova, E. (2009). A developmental approach to complex PTSD: Childhood and adult cumulative trauma as predictors of symptom complexity. *Journal of Traumatic Stress, 22*(5), 399–408. https://doi.org/10.1002/jts.20444

Creeden, K. (2009). How trauma and attachment can impact neurodevelopment: Informing our understanding and treatment of sexual behaviour problems. *Journal of Sexual Aggression, 15*(3), 261–273. https://doi.org/10.1080/13552600903335844

de Vries Robbé, M., de Vogel, V., Koster, K., & Bogaerts, S. (2015). Assessing protective factors for sexually violent offending with the SAPROF. *Sexual Abuse, 27*(1), 51–70. https://doi.org/10.1177/1079063214550168

Douglas, K. R., Chan, G., Gelernter, J., Arias, A. J., Anton, R. F., Weiss, R. D., Brady, K., Poling, J., Farrer, L., & Kranzler, H. R. (2010). Adverse childhood events as risk factors for substance dependence: Partial mediation by mood and anxiety disorders. *Addictive Behaviors, 35*(1), 7–13. https://doi.org/10.1016/j.addbeh.2009.07.004

Edwards, V. J., Holden, G. W., Felitti, V. J., & Anda, R. F. (2003). Relationship between multiple forms of childhood maltreatment and adult mental health in community respondents: Results from the adverse childhood experiences study. *American Journal of Psychiatry, 160*(8), 1453–1460. https://doi.org/10.1176/appi.ajp.160.8.1453

Finkelhor, D., & Kendall-Tackett, K. (1997). A developmental perspective on the childhood impact of crime, abuse, and violent victimisation. In D. Cicchetti & S. Toth (Eds.), *Rochester Symposium on Developmental Psychopathology: Developmental perspectives on trauma: Theory, Research, and Intervention* (Vol. 8, pp. 1–32). University of Rochester Press.

Finkelhor, D., Turner, H. A., Hamby, S. L., & Ormrod, R. (2011). *Polyvictimization: Children's exposure to multiple types of violence, crime, and abuse.* US Department of Justice, Office of Justice Programs, Office of Juvenile Justice and Delinquency Prevention.

Harris, D. A., & Levenson, J. S. (2020). Life on "the list" is a life lived in fear: Post-conviction traumatic stress in men convicted of sexual offenses. *International Journal of Offender Therapy and Comparative Criminology*. https://doi.org/10.1177/0306624x20952397

Herman, J. L. (1992). Complex PTSD: A syndrome in survivors of prolonged and repeated trauma. *Journal of Traumatic Stress, 5*(3), 377–391. https://doi.org/10.1002/jts.2490050305

Herman, J. L. (1997). *Trauma and recovery* (Vol. 551). Basic books.

Kruttschnitt, C., Uggen, C., & Shelton, K. (2000). Predictors of desistance among sex offenders: The interaction of formal and informal social controls. *Justice Quarterly, 17*(1), 61–88. https://doi.org/10.1080/07418820000094481

Levenson, J. S., & Morin, J. W. (2021). *The road to freedom* (4 ed.). Distributed by Safer Society Press.

Levenson, J. S., Willis, G., & Prescott, D. (2017). *Trauma-informed care: Transforming treatment for people who sexually abuse*. Safer Society Press.

Loper, A. B., Mahmoodzadegan, N., & Warren, J. I. (2008). Childhood maltreatment and cluster B personality pathology in female serious offenders. *Sexual Abuse: A Journal of Research and Treatment, 20*(2), 139–160. https://doi.org/10.1177/1079063208317463

Maschi, T., Baer, J., Morrissey, M. B., & Moreno, C. (2013). The aftermath of childhood trauma on late life mental and physical health a review of the literature. *Traumatology, 19*(1), 49–64. https://doi.org/10.1177/1534765612437377

Rutter, M., & Sroufe, L. (2000). Developmental psychopathology: Concepts and challenges. *Development and Psychopathology, 12*(03), 265–296. https://doi.org/10.1017/S0954579400003023

SAMHSA. (2014). *SAMHSA's concept of trauma and guidance for a trauma-informed approach*. Substance Abuse and Mental Health Services Administration.

Streeck-Fischer, A., & van der Kolk, B. A. (2000). Down will come baby, cradle and all: Diagnostic and therapeutic implications of chronic trauma on child development. *Australian and New Zealand Journal of Psychiatry, 34*(6), 903–918. https://doi.org/10.1080/000486700265

van der Kolk, B. (2005). Developmental trauma disorder: Toward a rational diagnosis for children with complex trauma histories. *Psychiatric Annals, 35*, 401–408. https://doi.org/10.3928/00485713-20050501-06

van der Kolk, B. (2006). Clinical implications of neuroscience research in PTSD. *Annals of the New York Academy of Sciences, 1071*(1), 277–293. https://doi.org/10.1196/annals.1364.022

van der Kolk, B. (2014). *The body keeps the score: Brain, mind, and body in the healing of trauma*. Penguin.

Weiss, M. J. S., & Wagner, S. H. (1998). What explains the negative consequences of adverse childhood experiences on adult health. *American Journal of Preventive Medicine, 14*(4), 356–360.

Yates, P. M., Prescott, D., & Ward, T. (2010). *Applying the good lives and self-regulation models to sex offender treatment: A practical guide for clinicians*. Safer Society Press

Young, J. E., Klosko, J. S., & Weishaar, M. E. (2003). *Schema therapy: A practitioner's guide*. Guilford Press.

The manufacturer's authorised representative in the EU is Springer Nature Customer Service Centre GmbH, Europaplatz 3, 69115 Heidelberg, Germany. If you have any concerns regarding our products, please contact ProductSafety@springernature.com

Printed and bound by CPI Group (UK) Ltd, Croydon, CR0 4YY
17/09/2026
02218454-0002